SCIENCE WORKSHOP SERIES

Biology

Annotated Teacher's Edition

LIFE PROCESSES

Seymour Rosen

GLOBE FEARON

Pearson Learning Group

CONTENTS

ISBN 0-130-23385-4 (Student Edition)

ISBN 0-130-23386-2 (Annotated Teacher's Edition

Printed in the United States of America

3 4 5 6 7 8 9 10   06 05 04 03 02

Formerly titled Dynamic Processes

INTRODUCTION TO THE SERIES

Overview

The *Science Workshop Series* consists of 12 softbound workbooks that provide a basic secondary-school science program for students achieving below grade level. General competency in the areas of biology, earth science, chemistry, and physical science is stressed. The series is designed so that the books may be used sequentially within or across each of these science areas.

Each workbook consists of approximately 30 lessons. Each lesson opens with a manageable amount of text for students to read. The succeeding pages contain exercises, many of which include photographs or drawings. The illustrations provide students with answers to simple questions. Phonetic spellings and simple definitions for scientific terms are also included to aid in the assimilation of new words.

The question material is varied and plentiful. Exercises such as *Completing Sentences, Matching,* and *True or False* are used to reinforce material covered in the lesson. An open-ended *Reaching Out* question often completes the lesson with a slightly more challenging, yet answerable question.

Easy-to-do laboratory experiments are also included in some lessons. Not isolated, the experiments are part of the development of concepts. They are practical experiments that require only easily obtainable, inexpensive materials.

Numerous illustrations and photographs play an important role in the development of concepts as well. The functional art enhances students' understanding and relates scientific concepts to students' daily lives.

The workbook format meets the needs of the reluctant student. The student is given a recognizable format, short lessons, and questions that are not overwhelming. The student can handle the stepwise sequence of question material in each lesson with confidence.

The series meets the needs of teachers as well. The workbooks can either be used for an entire class simultaneously or, since the lessons are self-contained, the books can be used on an individual basis for remedial purposes. This works well because the tone of each book is informal; a direct dialogue is established between the book and the student.

Using the Books

Although each lesson's reading selection may be read as part of a homework assignment, it will prove most effective if it is read during class time. This allows for an introduction to a new topic and a possible discussion. Teacher demonstrations that help to reinforce the ideas stressed in the reading should also be done during class time.

The developmental question material that follows the reading can serve as an ideal follow-up to do in class. The exercises such as *Completing Sentences, True or False,* or *Matching* might be assigned for homework or used as a short quiz. The *Reaching Out* question might also be assigned for homework or completed in class to end a topic with a discussion.

Objectives

The aim of the *Science Workshop Series* is to increase the student's level of competency in two areas: *Science Skills* and *Verbal Skills.* The comprehensive skills matrix on page T-2 highlights the science skills that are used in each lesson.

SKILLS MATRIX

Lesson	Identifying	Classifying	Observing	Measuring	Inferring	Interpreting	Predicting	Modeling	Experimenting	Organizing	Analyzing	Understanding Direct and Indirect Relationships	Inductive Reasoning	Deductive Reasoning
1	●		●								●			●
2		●			●		●			●			●	
3		●								●		●		●
4	●										●			
5			●	●				●	●				●	●
6			●				●				●	●		
7	●					●				●	●			●
8			●		●		●							
9			●		●		●			●		●		
10	●		●											●
11			●	●	●		●		●			●		
12		●				●					●		●	●
13		●				●	●	●	●		●	●	●	
14			●		●		●						●	
15	●				●		●					●		●
16		●	●							●				
17			●	●	●		●					●		
18		●								●				
19			●									●	●	
20	●		●		●		●							●
21		●						●		●	●	●	●	
22		●								●				
23			●					●			●			●
24			●	●	●	●	●			●			●	
25	●		●		●		●				●			

T–2

VERBAL SKILLS

An important objective of the *Science Workshop Series* is to give all students—even those with reading difficulties—a certain degree of science literacy. Reading science materials is often more difficult for poor readers because of its special vocabulary. Taking this into account, each new word is introduced, defined, used in context, and repeated in order to increase its familiarity. The development of the vocabulary word **chromosome** is traced below to illustrate the usage of science words in the text.

1. The word **chromosome** is first defined and spelled phonetically in Lesson 2 on page 8.
2. The *Fill in the Blank* exercise on page 11 requires students to use the word **chromosome** in context.
3. The *True or False* exercise on page 14 requires students to review the definition of **chromosome**.
4. The word is reintroduced and used throughout Lesson 6, which covers sex chromosomes.

This stepwise development allows students to gradually increase their working science vocabulary.

Other techniques used to familiarize students with a specialized vocabulary are less formal and allow the student to have fun while reinforcing what has been learned. Several varieties of word games are used. For example:

- A *Word Scramble* appears on page 33.
- A *Word Search* appears on page 78.
- A *Crossword Puzzle* appears on page 140.

LANGUAGE DIVERSE POPULATIONS

Students with limited English proficiency may encounter difficulties with the core material as well as the language. Teachers of these students need to use ample repetition, simple explanations of key concepts, and many concrete examples from the students' world. Relying on information students already possess helps students gain confidence and establishes a positive learning environment.

To help LDP students with language development, it is important to maintain an open dialogue between the students and the teacher. Encourage student participation. Have students submit written and oral reports. After students read a section of the text, have them explain it in their own words. These strategies will help the teacher be aware of problem areas.

CONCEPT DEVELOPMENT

In each book, the lessons are arranged in such a way as to provide a logical sequence that students can easily follow. Let us trace the development of one concept from the workbook: *Viruses cause diseases that affect humans.*

Lesson 15 familiarizes students with the characteristics of viruses. The term virus is defined and different kinds and shapes of viruses are explained. In Lessons 16 and 17, two types of disease—infectious and noninfectious—are defined and examples given. Lesson 16 emphasizes AIDS as an infectious disease. Lesson 18 deals with how the body protects itself from disease. In Lesson 19, ways of treating disease are explained.

SAFETY IN THE SCIENCE LABORATORY

Many of the lessons in the books of the *Science Workshop Series* include easy-to-do laboratory experiments. In order to have a safe laboratory environment, there are certain rules and procedures that must be followed. It is the responsibility of every science teacher to explain safety rules and procedures to students and to make sure that students adhere to these safety rules. Safety symbols also appear throughout the student's edition to alert students to potentially dangerous situations.

USING THE TEACHER'S EDITION

The Teacher's Edition of *Biology: Life Processes* has on-page annotations for all questions, exercises, charts, tables, and so on. It also includes front matter with teaching suggestions for each lesson in the book. Every lesson begins with questions to motivate the lesson. The motivational questions relate to the lesson opener pictures and provide a springboard for discussion of the lesson's science concepts. Following the *Motivation* are a variety of teaching strategies. Suggestions for *Class Activities, Demonstrations, Extensions, Reinforcements,* and *Cooperative/Collaborative Learning* opportunities are given.

If a student experiment is included in the lesson, a list of materials needed, safety precautions, and a short explanation of laboratory procedure are given.

The teacher's edition also includes a two-page test, which includes at least one question from each lesson in the book. The text can be photocopied and distributed to students. It begins on the next page. The test's Answer Key is found below.

ANSWER KEY

Multiple Choice

1. d   **2.** c   **3.** c   **4.** d   **5.** c   **6.** c   **7.** c   **8.** a   **9.** c

Fill In The Blank

1. genes   **2.** X, Y   **3.** adapted   **4.** fossils   **5.** antibodies   **6.** traits   **7.** hybrid
8. webs   **9.** grasses, weeds

True Or False

1. true   **2.** false   **3.** false   **4.** true   **5.** true   **6.** true   **7.** false   **8.** true

Matching

1. e   **2.** b   **3.** d   **4.** c   **5.** g   **6.** a   **7.** i   **8.** f   **9.** h

Name _____ Class _____ Date _____

MULTIPLE CHOICE

In the space provided, write the letter of the word or words that best complete each statement.

_____ 1. Ecology is the study of the environment and
   a) water. **b)** sunlight. **c)** soil. **d)** living things.

_____ 2. When one parent is pure dominant and the other is pure recessive, all the offspring are
   a) pure recessive. **b)** hybrid recessive. **c)** hybrid dominant.
   d) pure dominant.

_____ 3. The process by which organisms change over time is
   a) variation. **b)** natural selection. **c)** evolution. **d)** adaptation.

_____ 4. The viral disease that attacks a person's immune system is
   a) heart disease. **b)** cancer. **c)** measles. **d)** AIDS.

_____ 5. A food web shows the relationship between
   a) renewable resources. **b)** natural resources. **c)** food chains. **d)** producers.

_____ 6. A change in the environment does
   a) cause a change in genes. **b)** not cause mutations.
   c) not cause a change in genes. **d)** cause a change in DNA.

_____ 7. Any remain or trace of a once-living organism is a
   a) skeleton. **b)** shell. **c)** fossil. **d)** sediment.

_____ 8. Resistance to a specific disease is a(n)
   a) immunity. **b)** tumor. **c)** antibody. **d)** vaccine.

_____ 9. Plants are
   a) consumers. **b)** decomposers. **c)** producers. **d)** scavengers.

FILL IN THE BLANK

Complete each statement using a term from the list below. Write your answers in the spaces provided.

adapted	genes	traits	X
antibodies	grasses	webs	Y
fossils	hybrid	weeds	

1. A chromosome is made up of a chain of _____ .

2. There are two kinds of sex chromosomes, called _____ and

   _____ .

3. An organism that is suited to its environment is said to be _____ to its surroundings.

4. Most of what we know about human evolution has come from studying _____ .

Name _____ Class _____ Date _____

5. The immune system produces _____ .

6. The characteristics a living thing has are called _____ .

7. An individual that has both dominant and recessive genes for a trait is _____ for that trait.

8. Food chains combine to form food _____ .

9. If a forest burns down, _____ and _____ are the first to grow.

TRUE OR FALSE

In the space provided, write "true" if the sentence is true. Write "false" if the sentence is false.

_____ 1. A body cell has paired chromosomes.

_____ 2. Offspring inherit the acquired traits of the parents.

_____ 3. Embryology is the study of adult organisms.

_____ 4. Viruses do not ingest or digest food.

_____ 5. Some antibodies are made by bacteria.

_____ 6. All primates have flexible fingers and toes.

_____ 7. A human body cell has a total of 23 chromosomes.

_____ 8. New forms of DNA are made in genetic engineering.

MATCHING

Match each term in Column A with its description in Column B. Write the correct letter in the space provided.

	Column A		Column B
_____	1. Punnett square	a)	wise use of resources
_____	2. cross breeding	b)	mating related, but different breeds
_____	3. malignant tumor	c)	characterized by fever, muscle pain, paralysis
_____	4. polio	d)	harmful mass of cells
_____	5. algae	e)	used to show gene combinations
_____	6. conservation	f)	sudden change in genes
_____	7. dominant trait	g)	main producers in lakes and oceans
_____	8. mutation	h)	XX
_____	9. female	i)	trait that shows up in offspring

LESSON TEACHING STRATEGIES

LESSON 1

What are traits? (pp. 1–6)

Motivation   Refer students to the lesson opener picture on page 1 and ask the following questions:

1. What are the obvious traits of the organisms shown?

2. What are some of your traits?

Class Activity   Ask students if they look like other members of their families. Challenge students to explain their answers. Students probably will refer to characteristics such as eye color, hair color, skin color, and height. List these characteristics on the chalkboard and identify them as inherited traits. Define inherited traits as characteristics that are passed from parents to their offspring.

Demonstration   Bring a variety of pea pods to class to demonstrate some of the traits studied by Mendel. For example, bring in smooth and wrinkled seeds, and green and yellow pods. Have students observe each seed and pod and record their observations.

LESSON 2

What are chromosomes? (pp. 7–14)

Motivation   Refer students to the lesson opener picture on page 7 and ask the following questions:

1. What is shown in the picture?

2. What kind of organism do you think this is?

Reinforcement   Before beginning this lesson, review the processes of cell division and meiosis. Remind students that during cell division, each chromosome makes a copy of itself. Have students recall that meiosis is the special process by which gametes are formed. Ask students to compare the number of chromosomes in gametes and in body cells. (Gametes have only half the number of chromosomes as body cells.) Point out that during fertilization, two gametes combine and the original number of chromosomes is restored in the zygote. Emphasize that since chromosomes control heredity, the zygote receives traits from both parents.

Reinforcement   Students may have difficulty understanding the meaning and roles of genes and chromosomes. Stress that chromosomes are found in pairs in body cells, but singularly in sperm and egg cells. Genes are dark bands found in chromosomes that determine the traits of an organism.

LESSON 3

What are dominant and recessive traits? (pp. 15–20)

Motivation   Refer students to the lesson opener picture on page 15 and ask the following questions:

1. What is being shown in the pictures?

2. What traits do you have that are similar to traits of your parents?

Reinforcement   Be sure students understand that the lower case letter of the dominant trait is used to represent a recessive trait. Some students may be tempted to write the first letter of the recessive trait.

LESSON 4

How can we predict heredity? (pp. 21–26)

Motivation   Refer students to the lesson opener picture on page 21 and ask the following questions:

1. What is shown in the picture?

2. What other traits can be shown in a Punnett square?

Demonstration   Show students how Punnett squares are used to predict traits. Guide students through the examples shown on p. 22.

Class Activity   Have students practice monohybrid crosses using Punnett squares. Have students work in pairs or small groups. Place more able students with students of lesser ability to allow for peer tutoring.

Reinforcement   Draw Punnett squares on the chalkboard and diagram various monohybrid crosses. Identify the dominant trait and the recessive trait for each cross. Have students describe the genetic makeup and appearance of the offspring for each cross.

LESSON 5

What is incomplete dominance? (pp. 27–33)

Motivation Refer students to the lesson opener picture on page 27 and ask the following questions:

1. What is shown in the pictures?

2. What human traits are a result of incomplete dominance?

Demonstration Put blue food coloring in one cup of water and red food coloring in another. Mix the two cups of water together. The water will turn purple. Relate this activity to blending.

Reinforcement Emphasize that blended genes do not disappear. Draw a Punnett square on the chalkboard. Use the Punnett square to show students that blending occurs only in hybrids and that when hybrids are crossed, the pure traits appear again in their offspring.

Demonstration If possible, bring a white, a red, and a pink four-o'clock flower to class to show students an organism that exhibits blending.

LESSON 6

How is sex determined? (pp. 35–40)

Motivation Refer students to the lesson opener picture on page 35 and ask the following questions:

1. What is shown in the top picture?

2. What do the "X" and "Y" stand for in the bottom picture?

Discussion Refer students to the drawing of human chromosomes on page 36. Remind students that humans have 23 pairs of chromosomes. Have students observe pair 23 and describe the difference between the two chromosomes in the pair. Elicit the response that the two chromosomes are not identical. Point out that the chromosome map shown is that of a male. Explain that, in humans, males have different sex chromosomes called Y. Then describe how chromosomes in a sperm cell determine the sex of offspring. Be sure students understand that the sex chromosomes in females are identical.

Reinforcement Draw a Punnett square on the chalkboard showing the inheritance of sex. Ask students what ratio of offspring will be male (1/2) and what ratio will be female (1/2).

Reinforcement Be sure students understand that all body cells contain chromosome pair 23. Stress that sex chromosomes are not found only in gametes. Students often have this misconception.

Extension Have students research sex-linked traits. Tell students to write their findings in a report.

LESSON 7

How does the environment affect traits? (pp. 41–46)

Motivation Refer students to the lesson opener picture on page 41 and ask the following questions:

1. Why is the flower on the left healthy?

2. What is different about the flower on the right?

Demonstration Demonstrate how the environment might affect inherited traits. Show students two plants, one having received the proper amount of sunlight and one that was kept in a darkened place. Encourage students to describe the plants and to suggest how the unhealthy plant may be helped. Elicit the response that the plant should be put in sunlight. Point out that when the environment of a plant is not right, it will not grow and develop properly.

Reinforcement Emphasize that the genes for traits are not affected by the environment. Only the development of the trait is affected.

Extension Have interested students use library references to research the studies that have been done on identical twins to determine how the environment affects inherited traits. Have students present their findings in an oral report.

LESSON 8

What are some methods of plant and animal breeding? (pp. 47–52)

Motivation Refer students to the lesson opener picture on page 47 and ask the following questions:

1. What example of animal breeding is shown in the top picture?

2. What are the methods of mass selection?

Class Activity Bring in pictures showing many different breeds of dogs. Hold up each picture in front of the class. Have students identify as many breeds as they can. Point out that purebred animals are produced by inbreeding. Define inbreeding as the mating of closely related organisms.

Reinforcement When explaining hybridization, be sure students do not confuse organisms that are produced by hybridization, such as mules, with organisms that are hybrid for certain traits.

Extension If possible, arrange for a professional animal breeder to talk to your class about their work.

LESSON 9

What is genetic engineering? (pp. 53–58)

Motivation Refer students to the lesson opener picture on page 53 and ask the following questions:

1. What is shown in the two pictures?

2. What other kinds of fruit are a result of genetic engineering?

Demonstration Show students a model of gene splicing. Cut out a strip of construction paper. Bend the strip into a circle and staple the ends together to make a DNA chain. Then cut the chain open and insert a strip of different colored construction paper by stapling it to the original strip.

Reinforcement Be sure students understand that gene splicing is carried out by chemical means on large numbers of bacteria. Some students may envision a scientist snipping open a DNA chain and manually inserting genes.

Cooperative/Collaborative Learning Have a volunteer describe the three steps of gene splicing for the rest of the class.

LESSON 10

What is natural selection? (pp. 59–64)

Motivation Refer students to the lesson opener picture on page 59 and ask the following questions:

1. How does the animal represent natural selection?

Reinforcement Describe the four main ideas of Darwin's theory of evolution: overproduction, competition, variation, and survival of the fittest. Be sure students understand that evolution of a new species occurs over a long period of time.

Class Activity Bring in a variety of pictures showing members of the same species. Hold up each picture in front of the class. Tell students to observe how variations occur among members of the same species. Ask students to describe how the organisms shown in each picture are alike and how they are different.

Reinforcement Darwin's theory of evolution is often difficult for students to understand. Be sure students understand that the environment does not cause evolution. The environment only determines which variations will be selected.

Extension Have students research the work of Alfred Russell Wallace. Wallace was a British scientist who influenced Darwin's theory of evolution. Tell students to write their findings in a report.

LESSON 11

What evidence supports evolution? (pp. 65–72)

Motivation Refer students to the lesson opener picture on page 65 and ask the following questions:

1. What evidence of evolution is shown in the top picture?

2. What evidence of evolution is shown in the bottom picture?

Discussion Have students study Figure B on page 68 showing the evolutionary history of the horse. Emphasize that the fossil record of the horse is very complete. Compare the earliest horse to the modern horse. Point out the ways horses have changed over time. Tell students to note that the size of horses has changed. Ask students to describe other ways horses have changed. (The number of toes and the structure of the legs have changed.)

LESSON 12

How does adaptation help species survive? (pp. 73–78)

Motivation   Refer students to the lesson opener picture on page 73 and ask the following questions:

1. What adaptation does a camel have for living in its environment?

2. What adaptation does a penguin have for living in its environment?

Class Activity   Tell each student to carefully observe four different organisms and list as many different adaptations as they can for each of the organisms they observe. Review students' lists as a class. Have students explain how each organism's adaptations enable the organism to survive in its environment.

LESSON 13

What are the characteristics of primates? (pp. 79–84)

Motivation   Refer students to the lesson opener picture on page 79 and ask the following questions:

1. What primates are shown in the picture?

2. Are you a primate?

Discussion   Refer students to the pictures of hominid skulls on page 89. Have students compare the size of the skulls as you describe human evolution. Emphasize the trend toward increasing brain size.

Reinforcement   Draw a time line on the chalkboard highlighting human evolution.

LESSON 14

How did humans evolve? (pp. 85–93)

Motivation   Refer students to the lesson opener picture on page 85 and ask the following question:

1. What is shown in the picture?

Extension   Have interested students find out about one of the following hominid species: *Australopithecus afarensis, Australopithecus africanus, Homo habilis, Homo erectus.* Tell students to describe the characteristics of the species they choose in oral reports. Be sure students identify where, when, and by whom the species was discovered.

Cooperative/Collaborative Learning   Have one student describe the characteristics of Neanderthals for the rest of the class. Have another student describe Cro-magnons.

LESSON 15

What are viruses? (pp. 95–100)

Motivation   Refer students to the lesson opener picture on page 95 and ask the following questions:

1. Have you ever had a virus?

2. What are the three shapes of viruses?

Discussion   Have students review the definition of a living thing. Then discuss with students whether or not they think viruses are alive. Have students give reasons for their opinions. Write the main points of each argument on the board.

Class Activity   Have students sketch Figure C from page 98 on a sheet of white paper. Tell students to label the capsids and nucleic acids of the viruses. Remind students that all viruses are made up of nucleic acids and a capsid.

Discussion   Discuss the classification of viruses with students. Remind students that viruses are classified according to the living things they infect. Ask students to name other traits that are used to classify the animal viruses. (Students should mention the types of nucleic acids and capsids the viruses have.)

Extension   Discuss the various diseases and infections caused by viruses. Possible topics include pneumonia, influenza, the common cold, hepatitis, AIDS, polio, measles, and mumps. Ask students to describe what they know about these illnesses, such as how the virus spreads, if it is fatal, if a cure exists, and how to prevent or treat the infection. Write the facts about each virus on the board, while clearing up any misconceptions the students have.

LESSON 16

What are infectious diseases? (pp. 101–108)

Motivation   Refer students to the lesson opener picture on page 101 and ask the following questions:

1. What are some infectious diseases that you have had?

2. What is shown in the two pictures?

Extension   Have students bring to class recent newspaper and magazine articles about AIDS. Read and discuss some of the articles as a class.

Discussion   All students probably know that they should cover their nose and mouth when they sneeze or cough in order to prevent spreading germs. Use this familiar example to introduce the ways diseases are spread. Explain that when an infected person sneezes or coughs, substances that cause disease are spread into the air. Then discuss each of the other means of transmission: waterborne infections, foodborne infections, infections carried by insects or other vectors, and human carriers. Make sure students understand that human carriers who have no symptoms of the disease can transmit certain infectious diseases.

Class Activity   Have students look up the meaning of each work of the acronym AIDS: acquired; immune; deficiency; syndrome. Tell students to use the meanings of each word to write a brief description of AIDS in their own words.

LESSON 17

What are noninfectious diseases? (pp. 109–116)

Motivation   Refer students to the lesson opener picture on page 109 and ask the following questions:

1. What are some ways you can avoid getting a noninfectious disease?

2. What is shown in the pictures?

Class Activity   To introduce this lesson, write the factors contributing to heart disease on the chalkboard: age, family history, gender, smoking, high blood pressure, obesity, physical inactivity, and high cholesterol levels. Then ask a volunteer to go up to the chalkboard and circle any of the risk factors that can be controlled. Discuss the student's selections. Guide

students to understand that there are ways they can present heart disease. Have students list the names of several noninfectious diseases and state the causes of those diseases.

Class Activity   Have students list the names of several noninfectious diseases and state the causes of those diseases.

Extension   Have interested students research cancer or heart disease. Students should report on the history of the disease, prevention of the disease and how it is treated. Have students report their findings to the class.

Discussion   Describe how cancer spreads and some of its causes. Then refer students to the table on page 113 and discuss the warning signs of cancer.

Cooperative/Collaborative Learning   Have a volunteer explain the difference between a benign tumor and a malignant tumor for the rest of the class.

LESSON 18

How does the body protect itself from disease? (pp. 117–124)

Motivation   Refer students to the lesson opener picture on page 117 and ask the following questions:

1. What is being shown in the picture on the left?

2. How do you think the skin protects the body from disease?

Demonstration   To show students that microorganisms are everywhere, perform the following activity: Make up two petri dishes of nutrient agar. Leave one dish of sterile agar exposed to the air for 20 minutes. Leave the other dish closed. Place both dishes in a dark, warm place for several days. After several days, have students observe the petri dishes and discuss the results. (Bacterial colonies will grow in the dish that was exposed to the air; no bacteria will grow in the control dish.) Then sterilize and discard the dishes properly.

Discussion   Ask students to describe the function of a country's military. (defense) Have students name different branches of the military. (army, navy, air force, marines, and so on) Explain that the human body is constantly waging a battle against substances that cause disease. Just as most countries have different military branches for defense, so the body has different lines of defense to fight the substances that cause disease. Then describe the defenses of the human body.

Reinforcement   Ask students to recall the functions of white blood cells, lymph nodes, and the cilia and mucus found in the respiratory system.

LESSON 19

What are other ways of fighting disease? (pp. 125–128)

Motivation   Refer students to the lesson opener picture on page 125 and ask the following questions:

1. What is being shown in the two pictures?

2. How do you fight disease?

Reinforcement   Be sure students understand that a person can become allergic to a given antibiotic at any time in their lifetime. Also, once a person does experience an allergic reaction to a given antibiotic, such as penicillin, it is likely that the allergic reaction will worsen if that person takes the antibiotic again in the future.

Class Activity   Have students look up the meanings of the prefix *anti-* and the term *biotic* in a dictionary. (*anti-*: against; *biotic*: pertaining to life) Then relate the meanings of these words to the definition of antibiotic. Point out that antibiotics are substances that destroy bacteria.

Demonstration   If possible, show students a sample of *Penicillium notatum,* which is the mold most commonly used in penicillin production.

Reinforcement   Be sure students understand that some antibiotics are produced by some types of bacteria and used to destroy other bacteria, and that antibiotics do not work against viruses.

Reinforcement   Emphasize the importance of preventive medicine. Stress that decisions students make today relative to a healthy lifestyle will affect their futures.

LESSON 20

What is ecology? (pp. 129–132)

Motivation   Refer students to the lesson opener picture on page 129 and ask the following questions:

1. What is happening in the top picture that could harm the environment?

2. What are some of the living things shown in the bottom picture?

Extension   You may wish to introduce the terms biotic and abiotic. The biotic factors in an environment are the living things. Biotic factors include plants, animals, protists, bacteria, and fungi. Abiotic factors are the nonliving parts of the environment. Ask students to list as many abiotic factors as they can for their classroom environment. Write student responses on the chalkboard.

LESSON 21

What are some other characteristics of an ecosystem? (pp. 133–140)

Motivation   Refer students to the lesson opener picture on page 133 and ask the following question:

1. What is being shown in the picture?

Reinforcement   Hold up a nature scene showing a variety of different kinds of animals. Ask students how many different animal populations are shown. Tell students that all the populations together make up a community. Then point out some of the nonliving parts of the environment in the picture. Ask students what the living and nonliving parts of the environment make up. (ecosystem)

LESSON 22

What are biomes? (pp. 141–146)

Motivation   Refer students to the lesson opener picture on page 141 and ask the following questions:

1. What biome do you live in?

2. What animals and vegetation are prevalent where you live?

Reinforcement Be sure students understand why organisms may have the same habitat but not the same niche. Emphasize that competition among species living in the same place is the reason organisms cannot share the same niche.

Extension You may wish to tell students that animals that eat only plants are called herbivores, animals that eat only meat are called carnivores, and animals that eat both plants and animals are called omnivores.

Class Activity Have students work in small groups to construct various food chains and food webs. Tell students to label the producers and the consumers on their models. Have students illustrate their models with pictures. When all groups are finished, have a representative from each group display and describe the group's models for the rest of the class.

Reinforcement Refer students to the biome map on page 143. Have students locate the six major biomes as you describe the plants and animals that are characteristics of each biome.

Extension Have interested students research the marine biome which is the largest biome of Earth. Tell students to find out the characteristics of the ocean, estuary, and intertidal zone. Tell students to write their findings in a report.

LESSON 23

What things can change the environment? (pp. 147–152)

Motivation Refer students to the lesson opener picture on page 147 and ask the following question:

1. What things do you think change the environment?

Extension Succession can occur when an ecosystem is changed due to fire, wind, farming, and so on. An ecosystem also can change when a new habitat, such as an island, is created. Ecological succession also can start in areas that did not previously support life. Have students research how lichens can start ecological succession. Tell students to write their findings in a report.

LESSON 24

How do people upset the balance of nature? (pp. 153–157)

Motivation Refer students to the lesson opener picture on page 153 and ask the following questions:

1. What is being shown in the three pictures?

2. How can you prevent the upset of the balance of nature?

Class Activity Divide the class into four groups. Have each group draw one of the following communities on a posterboard; open field, shrub land, pine forest, hardwood forest. Then display the student's artwork in sequence to illustrate succession.

LESSON 25

What is conservation? (pp. 159–166)

Motivation Refer students to the lesson opener picture on page 159 and ask the following questions:

1. What is shown in the two pictures?

2. How do you conserve resources?

Class Activity Write the term *Pollution* on the chalkboard. Discuss what students think of when they hear the word *pollution*. Emphasize that pollution is upsetting the balance of the environment.

Extension Have students find out what measures are being taken to prevent endangered animals from becoming extinct. Tell students to present their findings in an oral report.

Extension Have students read Rachel Carson's *Silent Spring* and write a book report.

SCIENCE WORKSHOP SERIES

Biology

LIFE PROCESSES

Seymour Rosen

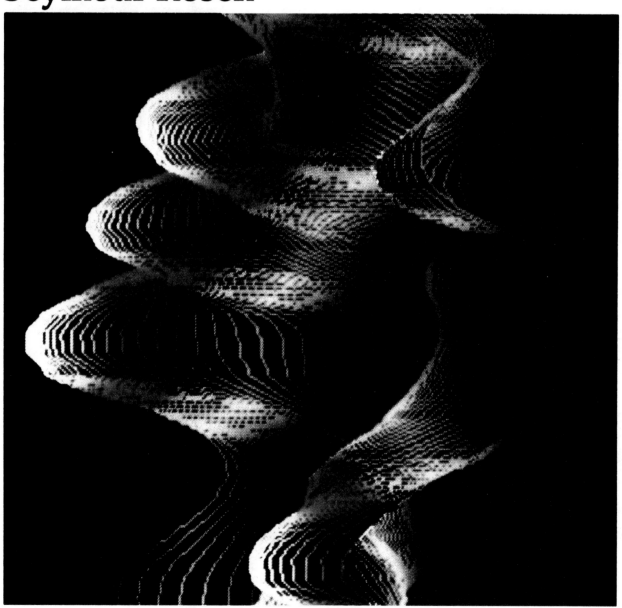

GLOBE FEARON

Pearson Learning Group

THE AUTHOR

Seymour Rosen received his B.A. and M.S. degrees from Brooklyn College. He taught science in the New York City School System for twenty-seven years. Mr. Rosen was also a contributing participant in a teacher-training program for the development of science curriculum for the New York City Board of Education.

Cover Designer: Joan Jacobus
Cover Photograph: Lawrence Berkeley Laboratory/Photo Researchers, Inc.
Cover Photo Researcher: Joan Jacobus
Photo Researchers: Rhoda Sidney, Jenifer Hixson

About the cover illustration: This computer image of DNA (deoxyribonycleic acid) shows the spiral arrangement of the DNA molecules that carry a person's genes.

Photo Credits:

p. 10, Fig. D: Runk/Schoenberger, Grant Heilman
p. 34: Alan Carey/The Image Works
p. 61, Fig. A: National Audubon Society/Photo Researchers
p. 61, Fig. B: Hal Harrison/Grant Heilman
p. 62, Fig. C: Christian Grzimek/Photo Researchers
p. 62, Fig. D: National Audubon Society/Photo Researchers
p. 82, Fig. C: Peter Menzel/Stock, Boston
p. 94: Heilman/Monkmeyer Press Photos
p. 115, Fig. H: PhotoTrends
p. 116, Fig. I: Rhoda Sidney
p. 123, Fig. F: Communicable Disease Center
p. 123, Fig. G: Christopher Morrow/Stock, Boston
p. 144, Fig. B: Grant Heilman
p. 144, Fig. C: National Audubon Society/Photo Researchers
p. 144, Fig. D: Carl Frank, Photo Researchers
p. 144, Fig. E: Lee Snider, Photo Images
p. 145, Fig. F: Omikron/Photo Researchers
p. 145, Fig. G: Fredrik D. Bodin/Stock, Boston
p. 155, Fig. A: United Nations
p. 155, Fig. B: Runk/Schoenberger/Grant Heilman
p. 158: Peter Menzel/Stock, Boston
p. 161, Fig. A: Bethlehem Steel Corporation
p. 162, Fig. C: Photo Researchers
p. 163, Fig. E: Photo Researchers
p. 163, Fig. F: Advertising Council, U.S.D.A. Forest Service
p. 164, Fig. G: Grant Heilman
p. 164, Fig. H: Wendell Metzen/Bruce Coleman

ISBN: 0-130-23385-4 (Student Edition)

Printed in the United States of America

3 4 5 6 7 8 9 10 04 03 02

Formerly titled Dynamic Processes

1-800-321-3106
www.pearsonlearning.com

CONTENTS

Introduction to Life Processes

Has anyone ever told you that you have your mother's eyes? What did they mean? In this book, you will learn about the close similarities between parents and their offspring. Offspring resemble their parents because they inherit certain traits, or characteristics from them. You will learn the reasons why you may resemble either of your parents.

You also will learn about evolution, or the process by which organisms change over time. We will trace the path of theories of evolution from the time of Charles Darwin to present-day theories.

In this book, you also will learn about viruses and disease. The difference between infectious disease and noninfectious disease will be explained. In addition, ways of preventing diseases, such as AIDS, will be discussed.

Finally, you will learn more about ecology and conservation and what you can do to make the environment a better place in which to live.

What are traits?

KEY TERM

traits: characteristics of living things

LESSON 1 | What are traits?

It is easy to recognize an elephant. An elephant is very large and has a long trunk. A giraffe is easy to recognize too—by its long neck.

The elephant's trunk and the giraffe's neck are examples of **traits**. Traits are characteristics that living things have. They help us to identify living things.

Scientists have divided living things into groups according to traits. All members of a group have certain traits that are the same. For example, all birds have feathers. All mammals have some hair. All giraffes have long necks. And all elephants are large and have long trunks.

Organisms within a group may share certain traits, but no two are exactly alike. There are always <u>individual</u> differences. We call these differences <u>individual traits</u>.

Take the elephant for example. All elephants are large, but some are larger than others. All giraffes have long necks, but some giraffes have longer necks than others.

All humans share certain traits. However, no two people are exactly alike—not even identical twins. There are always individual differences.

Individual differences enable us to identify different members of the same group.

Think of your friends, for example. You know one from another by their individual traits. They include differences in size, hair type and coloring, skin coloring, and shape of face. How many other human traits can you name?

Figure A

Humans and frogs are alike in **some ways.** They share certain traits. **For example:**

- Both humans and frogs **are living** things. Therefore, **both carry out the** life processes.

- Both humans and frogs **are animals.**

- Both humans and frogs are **vertebrates.** They have backbones.

But humans and frogs are different from one another too—very different. **We can tell** humans from frogs by the traits they do not share.

Fifteen traits are listed below and on the next page. Some are **human traits. Some are traits** that frogs have.

*Study each trait. Does it belong to humans or does it belong to frogs? Write **Human** next to each human trait. Write **Frog** next to each frog trait.*

1. some hair covering _____ Human _____

2. external fertilization _____ Frog _____

3. internal fertilization _____ Human _____

4. embryos develop outside the female's body _____ Frog _____

5. females can nurse their young _____ Human _____

6. give birth to live young _____ Human _____

7. live entire life on land _____ Human _____

8. live early part of life in water and adult life on land _____ Frog _____

9. breathe by lungs only _____ Human _____

10. breathe through gills in early life _____ Frog _____

11. adults breathe by lungs or through skin _____ Frog _____

12. stand on two legs _____ Human _____

13. stand on four legs _____Frog_____

14. eat mostly insects _____Frog_____

15. eat meat and plants _____Human_____

Now answer these questions.

16. Do all humans have the traits you have listed as "Human"? _____yes_____

17. Do all frogs have the traits you have listed as "Frog"? _____yes_____

18. The traits you have listed are all _____group_____ traits.

group, individual

19. Are all frogs exactly alike? _____no_____

20. Are all humans exactly alike? _____no_____

IDENTIFYING INDIVIDUALS BY INDIVIDUAL TRAITS

Look at Figure B and then answer the questions.

John, Jim, and Tom are humans. They are about the same age. They have all the traits that humans share. Yet, they are different from one another.

Figure B

- John is short and thin. He has light-brown skin and dark straight hair.

- Jim is tall and heavy. He has dark-brown skin and dark curly hair.

- Tom is tall and thin. He has fair skin and light curly hair.

1. Identify by letter.

   a) Which one is John? _____b_____

   b) Which one is Jim? _____a_____

   c) Which one is Tom? _____c_____

2. a) Do all humans have hair? _____yes_____

   b) Hair _____is_____ a human trait.
        is, is not

   c) Do all humans have the same color hair? _____no_____

   d) Do all humans have curly hair? _____no_____

   e) Do all humans have straight hair? _____no_____

3. What kind of trait is hair color and type? _____individual_____
                                                    individual, group

4. At a given age, is every person the same height? _____no_____

5. Are some people taller than average? _____yes_____

6. Are some people shorter than average? _____yes_____

7. Difference in height is what kind of trait? _____individual_____
                                                     individual, group

Figure C shows two pea pods and their peas. Both are the same age.

8. Are the peas of these pods exactly

   the same? _____no_____

9. What difference do you notice in the skins of the peas?

   _____some are smooth; some are_____

   _____wrinkled_____

Figure C

10. What kind of difference is this? _____individual_____
                                          individual, group

Complete each statement using a term or terms from the list below. Write your answers in the spaces provided.

group   the same   plants
traits   individual traits exactly
humans   individual   identified
living things

1. The characteristics a living thing has are called _____traits_____ .

2. Living things are _____identified_____ by their traits.

3. Scientists group _____living things_____ according to their traits.

4. Members of a group have certain traits that are _____the same_____ .

5. No two living things are _____exactly_____ the same.

6. Differences among individuals of the same group are called _____individual traits_____ .

7. Having a spinal cord, internal fertilization and embryo development are group

    traits of _____humans_____ .

8. Individual differences enable us to identify different members of the same

    _____group_____ .

9. Having cell walls and making their own food are group traits of

    _____plants_____ .

10. Wrinkled skin or smooth skin are _____individual_____ traits of peas.

Match each term in Column A with its description in Column B. Write the correct letter in the space provided.

	Column A	Column B
__d__	1. traits	**a)** group trait of apple trees
__a__	2. green leaves	**b)** individual human trait
__e__	3. number of green leaves	**c)** human group trait
__c__	4. hair	**d)** characteristics
__b__	5. hair texture	**e)** individual trait of apple trees

What are chromosomes?

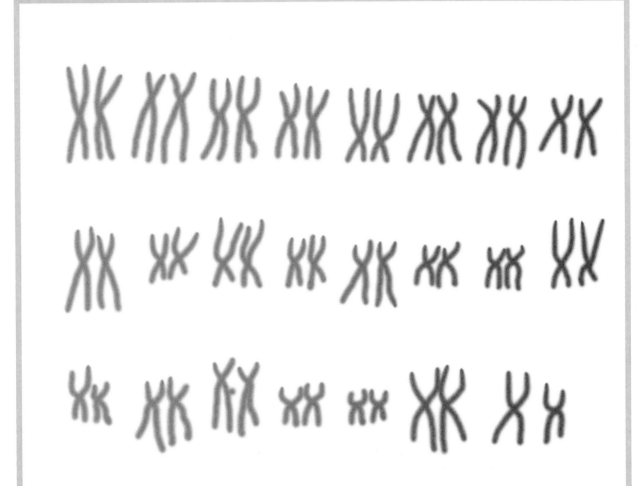

KEY TERMS

chromosome: threadlike structures in the nucleus of a cell that control **heredity**

gene: part of a chromosome that controls inherited traits

gamete: sex cell

genetics: study of heredity

LESSON 2 | What are chromosomes?

"Mary has her mother's eyes." "Tom is built just like his father." How often have you heard remarks like these?

All people resemble their parents in some ways. They have similar traits. . . . And it is no accident. Many traits are passed on from parents to offspring. We say they are <u>inherited</u>. How are they inherited? The answer is found in the cell nucleus.

The nucleus has tiny bodies called **chromosomes** [KROH-muh-sohms]. Most are rod-shaped. In body cells, chromosomes are found in pairs. Body cells are all the cells <u>except</u> sperm and egg cells.

Each kind of organism has a specific number of chromosomes. For example, every body cell of a fruit fly has 8 chromosomes (4 pairs); a human has 46 (23 pairs); a garden pea has 14 (7 pairs).

Along each chromosome there are many dark bands. Each band is a small part of a chromosome called a **gene**. There are many, many genes, at least one million in every nucleus. <u>Genes determine the traits of an organism.</u>

There are genes for height, genes for nose size and shape, genes for the color of hair, skin, and eyes. In fact, there are genes for most traits any individual has. Some genes even affect traits like voice, intelligence, and behavior. Genes also control the life processes of your cells.

In both asexual and sexual reproduction, chromosomes (and genes) are passed from parents to offspring. During asexual reproduction, each daughter cell receives chromosomes from a single parent cell. The daughter cell is an exact copy of the parent. Some organisms and the body cells of <u>all</u> organisms reproduce asexually.

During sexual reproduction, an offspring receives chromosomes from each parent cell. The chromosomes in **gametes**, or sex cells, are <u>not</u> paired. A sperm or an egg cell has only half the number of chromosomes as a body cell. When fertilization takes place, the sperm cell and the egg cell unite. Together, their chromosomes add up to the full number of chromosomes found in body cells. The fertilized egg, or zygote, has chromosomes from both of its parents. It also has traits from both parents.

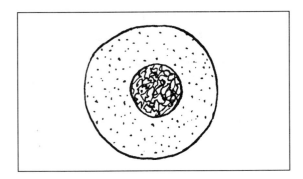

Figure A

Every cell has a nucleus.

1. Figure A shows an animal cell.

   a) Draw a line to the nucleus.

   b) Label it "nucleus."

   Check students' labels.

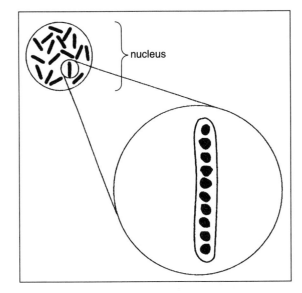

Figure B

2. A nucleus contains tiny rod-shaped bodies. What are they called?

   chromosomes

3. A chromosome is made up of even smaller bodies. What are they called?

   genes

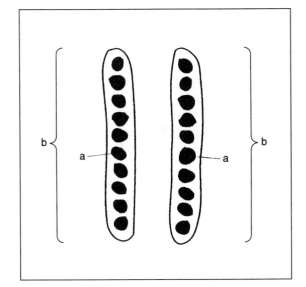

Figure C

4. Figure C shows a pair of chromosomes and their genes.

   a) The chromosomes are labeled

   b .

   b) Two genes are labeled ____a____ .

5. Why are genes important? _____

   They control traits.

9

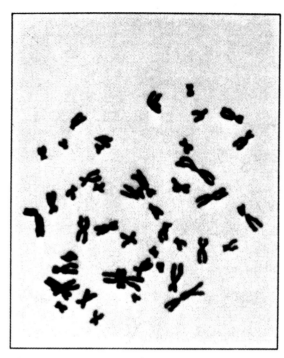

Figure D shows what actual human chromosomes look like.

- Every body cell of a particular organism has the same chromosomes.

- No two individuals that reproduce sexually have the same chromosomes.

You have trillions of body cells. Each cell has the same chromosomes. No one else in the world has the same chromosomes. There is no "duplicate" of you—anywhere!

Figure D *Human chromosomes*

The study of traits and how they are passed on is called **genetics** [juh-NET-iks].

- All living things have traits.

- All living things have genes.

- Only living things have genes.

Genes contain the "plans" for the traits an organism has.

What are genes made of? Scientists have discovered that genes are made of a complicated compound called DNA. DNA stands for deoxyribonucleic [dee-oks-ee-ry-boh-noo-KLEE-ik] acid. Try to pronounce it.

Figure E

Complete each statement using a term or terms from the list below. Write your answers in the spaces provided. Some words may be used more than once.

genes	46	specific
pairs	genetics	23
inherited	traits	chromosomes

1. The characteristics an individual has are called _____ traits _____.

2. Traits are passed down from parents to offspring. Another way of saying this is "traits are _____ inherited _____."

3. The study of heredity is called _____ genetics _____.

4. The nucleus has tiny rod-shaped bodies called _____ chromosomes _____.

5. A chromosome is made up of a chain of _____ genes _____.

6. Genes determine the _____ traits _____ of an individual.

7. Every organism has a _____ specific _____ number of chromosomes.

8. In body cells, chromosomes are found in _____ pairs _____.

9. Each of your body cells has _____ 23 _____ pairs of chromosomes. This is a total of _____ 46 _____ single chromosomes.

10. A human sperm or egg has _____ 23 _____ single chromosomes.

Match each term in Column A with its description in Column B. Write the correct letter in the space provided.

	Column A	Column B
__d__	1. genes	a) compound that makes up genes
__b__	2. chromosomes	b) made up of many genes
__a__	3. DNA	c) have unpaired chromosomes
__e__	4. body cells	d) pass on traits
__c__	5. gametes	e) have paired chromosomes

The pictures below show how chromosomes are passed from parent to offspring during asexual and sexual reproduction. Study Figures F and G. Then answer the questions.

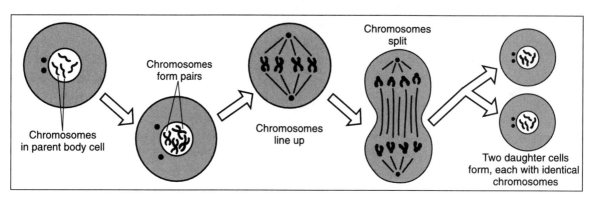

Figure F *Asexual reproduction*

1. How many chromosomes does the parent cell in Figure F have? _____4_____

2. How many chromosomes does each daughter cell have? _____4_____

3. In Figure F, how do the parent cell's chromosomes compare to the daughter cell's

   chromosomes? ____They are exactly the same._____

4. Which Figure shows how body cells reproduce? _____F_____

F, G

5. **a.** In Figure G, how many chromosomes does each sperm cell contain? _____4_____

   b. How many chromosomes does each egg cell contain? _____4_____

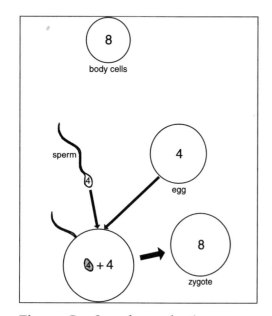

Figure G *Sexual reproduction*

6. Gametes have _____half_____ the

half, twice
   number of chromosomes as body cells.

7. Fertilization produces a single cell. What is

   it called? _____zygote_____

8. How many chromosomes does the zygote

   in Figure G have? _____8_____

9. How many chromosomes will each body cell

   of the organism have? _____8_____

10. The offspring will have traits of both the

    mother and father. Why? __It inherited genes__

    __from both parents. Genes control traits.__

12

Fill in the missing number of chromosomes.

	Organism	Chromosomes in each body cell	Chromosomes in each sperm or egg
1.	Human	46	23
2.	Horse	60	30
3.	Housefly	12	6
4.	Dog	78	39
5.	Grasshopper	14	7
6.	Mosquito	6	3
7.	Chicken	18	9
8.	Apple	34	17
9.	Spinach	12	6
10.	Lily	24	12

11. A gamete has _____half_____ the number of chromosomes that a body cell has.

half, twice

12. How many pairs of chromosomes are there in each body cell of the following?

a) horse ___30___

d) lily ___12___

b) mosquito ___3___

e) human ___23___

c) spinach ___6___

f) housefly ___6___

Below are several scrambled words you have used in this Lesson. Unscramble the words and write your answers in the spaces provided.

1. NEEG _____ GENE _____

2. HERINIT _____ INHERIT _____

3. NEGITECS _____ GENETICS _____

4. ETEMAG _____ GAMETE _____

5. CHOMEOSORM _____ CHROMOSOME _____

TRUE OR FALSE

In the space provided, write "true" if the sentence is true. Write "false" if the sentence is false.

True   **1.** Traits are the characteristics of living things.

False   **2.** Only animals have traits.

False   **3.** Traits are passed on from offspring to parents.

True   **4.** Traits are passed on by genes.

False   **5.** A cell has only a few genes.

False   **6.** Only animals have genes.

True   **7.** Different genes control different traits.

True   **8.** Genes form chromosomes.

False   **9.** Every organism has the same number of chromosomes.

True   **10.** Body cells have paired chromosomes.

False   **11.** Gametes have paired chromosomes.

False   **12.** A body cell and a sex cell have the same number of chromosomes.

True   **13.** Gametes have half the number of chromosomes of body cells.

False   **14.** A human body cell has a total of 23 chromosomes.

True   **15.** A human gamete has 23 single chromosomes.

REACHING OUT

Which organism would more closely resemble its parent, one produced by asexual reproduction, or one produced by sexual reproduction? Why? _Answers will vary._

Possible student responses include: The organism produced by asexual reproduction

because it would receive all its chromosomes from one parent cell.

What are dominant and recessive traits?

KEY TERMS

dominant gene: stronger gene that always shows itself

recessive gene: weaker gene that is hidden when the dominant gene is present

pure: having two like genes

hybrid: having two unlike genes

LESSON 3 | What are dominant and recessive traits?

Tom has dark hair, just like his parents. Sally's hair is dark too, just like her father. Her mother's hair, however, is blonde.

It is easy to understand why Tom's hair is dark. Both of his parents have dark hair. How about Sally? Why is her hair dark? Why not blonde?

This kind of question was first answered in the mid 1800s by Gregor Mendel, an Austrian monk. Mendel often is called the "Father of Genetics." Mendel observed inherited traits. He wondered why certain traits found in parents show up in their offspring, while other traits do not.

To find the answer, Mendel experimented with pea plants. He observed certain traits such as tallness and shortness, color, and the smoothness of the seed coverings. His experiments led to the Principles of Genetics. These principles hold true for all organisms that reproduce sexually.

One of the principles of genetics is called the Law of Dominance. The Law of Dominance states:

1. An organism receives two genes for each trait, one from each parent.

2. One of the genes may be stronger than the other. The trait of the stronger gene is expressed, or shows up. The gene that shows up is called the **dominant** [DOM-uh-nunt] **gene**. The "hidden" gene is called the **recessive** [ri-SES-iv] **gene** for that trait.

If an offspring receives two of the same gene (either two recessive or two dominant), the offspring will inherit that trait. There is no other possibility.

However, suppose an organism has one dominant gene and one recessive gene for a certain trait. The organism will have the trait of the dominant gene. The recessive gene will be "hidden."

Let's look at Sally again. Sally has genes for dark hair and for light hair. The gene for dark hair is dominant over the gene for light hair. That is why Sally's hair is dark.

It is interesting to note that a trait that is dominant for one kind of organism may be recessive in another organism.

What You Need to Know:

Organisms that have two of the same genes for a certain trait are called **pure**.

A pure organism may have two dominant genes or two recessive genes. For example, a pea plant may have two genes for tallness or two genes for shortness. In pea plants, the gene for tallness is dominant.

Organisms that have two unlike genes for a certain trait are called **hybrid** [HY-brid]. A pea plant that has one gene for tallness and one gene for shortness is a hybrid.

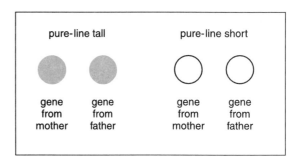

Figure A

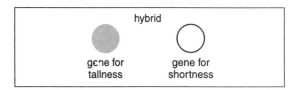

Figure B

No organism has all dominant or all recessive genes.

An organism may be pure in certain traits and hybrid in others. Figures C through F show some of Mendel's experiments with pea plants. Study the figures and answer the questions with each.

Circle the letter of the phrase that completes each sentence best. Fill in the answer blanks for the other sentences.

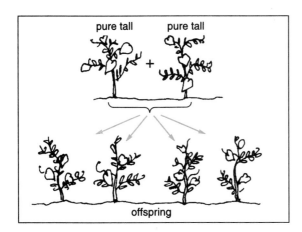

Figure C *Mendel cross-pollinated two pure tall pea plants.*

1. Offspring of pure tall pea plants are

   a) only tall.

   b) only short.

   c) tall and short.

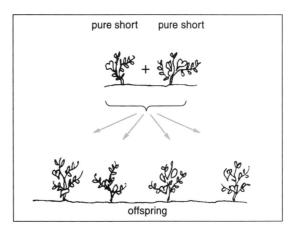

Figure D *Mendel crossed two pure short pea plants.*

2. Offspring of pure short pea plants are

   a) only tall.

   b) only short.

   c) tall and short.

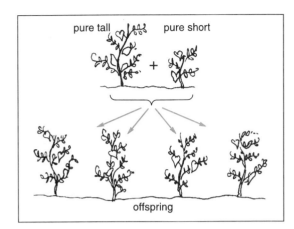

Figure E *Mendel crossed a pure tall with a pure short.*

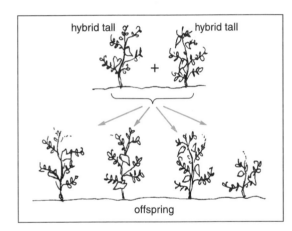

Figure F *Mendel crossed hybrid plants.*

3. Offspring of pure tall pea plants and pure short plants are

   a) only tall.

   b) only short.

   c) short and tall.

4. We see that in pea plants, ___tallness___

   shortness, tallness

   is dominant over ___shortness___ .

   shortness, tallness

5. The offspring now carry genes of height from both parents. They are

   a) genes only for tallness.

   b) genes only for shortness.

   c) genes for tallness and shortness.

6. The offspring are ___hybrids___ .

   pure, hybrids

7. Refer to Figure F. Offspring of hybrid-tall pea plants are

   a) only tall.

   b) only short.

   c) short and tall.

8. **a)** Which is the dominant trait? ___tallness___

   b) Does the dominant trait show up in every offspring? ___no___

   Look at Figure F.

9. **a)** Which trait is recessive? ___shortness___

   b) Is the recessive trait always hidden? ___no___

   c) How many plants are tall? ___3___

   d) How many plants are short? ___1___

10. Complete the fractions in these sentences:

    When you cross hybrids, the dominant trait shows up $\frac{3}{4}$ of the time.

    The recessive trait shows up $\frac{1}{4}$ of the time.

How many traits do you recognize in yourself?

Dominant	Recessive
brown eyes	blue eyes
very curly hair	wavy hair
wavy hair	straight hair
freckles	no freckles
nearsightedness	normal eyesight
long eyelashes	short eyelashes
large ears	small ears
dimpled cheeks	no dimples

PREDICTING HUMAN TRAITS

Now use the information from the chart above to fill in the chart below. The first example has been done for you.

	Mother	Father	Offspring	Dominant or Recessive?	Hybrid or Pure?
1.	normal eyes	nearsighted	nearsighted	dominant	hybrid
2.	straight hair	straight hair	straight hair	recessive	pure
3.	long eyelashes	short eyelashes	long eyelashes	dominant	hybrid
4.	no freckles	no freckles	no freckles	recessive	pure
5.	no dimples	dimples	dimples	dominant	hybrid
6.	blue eyes	brown eyes	brown eyes	dominant	hybrid
7.	large ears	large ears	large ears	dominant	pure
8.	wavy hair	very curly hair	very curly hair	dominant	hybrid

Now answer these questions.

9. How many offspring in the chart will be pure recessive for a trait? _____2_____

10. Why will the recessive genes show up? ___If an offspring receives two of the same genes, the offspring inherits that trait.___

Complete each statement using a term or terms from the list below. Write your answers in the spaces provided. Some words may be used more than once.

hybrid	Gregor Mendel	dominant
recessive	genes	pure
pea plants	are the same	two

1. A pioneer in the study of heredity was __Gregor Mendel__ .

2. Mendel studied heredity by experimenting with __pea plants__ .

3. Traits are controlled by __genes__ .

4. In organisms that reproduce sexually, every trait has genes from __two__ parents.

5. The "stronger" of the two traits which show up in an organism is called the __dominant__ trait.

6. The "weaker" of the two traits is called the __recessive__ trait.

7., No organism has all __dominant__ or all __recessive__ genes.

8. An organism whose genes for a trait are the same is called __pure__ for that trait.

9. An organism whose genes for a trait are not the same is called __hybrid__ for that trait.

10. An offspring will definitely inherit a trait if both its genes for that trait __are the same__ .

Match each term in Column A with its description in Column B. Write the correct letter in the space provided.

	Column A	**Column B**
__b__ 1.	dominant trait	**a)** has mixed genes for a given trait
__e__ 2.	recessive trait	**b)** shows up in offspring
__d__ 3.	pure	**c)** a dominant trait in pea plants
__a__ 4.	hybrid	**d)** has two like genes for a given trait
__c__ 5.	tallness	**e)** may remain "hidden"

How can we predict heredity?

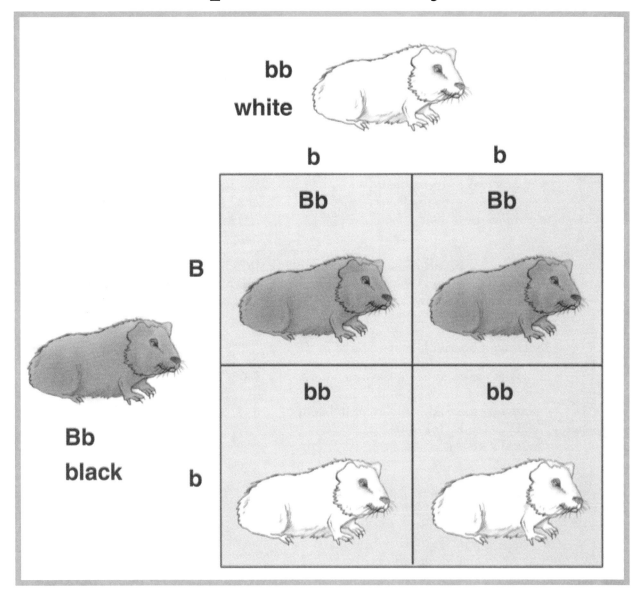

KEY TERM

Punnett square: chart used to show possible gene combinations

LESSON 4 | How can we predict heredity?

Meet Mr. and Mrs. Jones:

TOM   and      SUSAN

Tom has two of the same genes for hair color. He is pure for dark hair.

Susan has one **dominant** gene for dark hair **and one** recessive gene for **blonde hair. She is hybrid.**

How can we predict what color hair their children will have? **It's easy!** We can use a special chart called a **Punnett square**. A Punnett **square is** a chart used to show possible gene combinations. The steps **here show** you how to use a Punnett square.

1. Draw a box with four squares in it.

2. Write the genes from the father across the top of the chart. A dominant gene always marked with a capital letter "D" stands for dark hair. Both of Tom's genes for hair color are represented by D.

3. Write the genes from the mother down the side of the chart. A recessive gene is always marked with a lower case letter. Susan is hybrid dark for hair color. One gene is marked D. The other gene is the recessive gene for blond hair. The symbol for this gene is d.

4. Now fill in each box with a gene from the father and a gene from the mother. Each box now shows the different combination of genes that can show up in the offspring.

male gametes

	D	D
D		
d		

female gametes

	D	D
D	DD	DD
d	Dd	Dd

Look again at the Punnett square on page 22. What do the letters in the box tell us?

- The possible gene combination are DD, DD, Dd, and Dd.
- Each combination has a dominant gene for dark hair.
- Therefore all of Tom and Susan's children will have dark hair.

If Tom and Susan have four children, the Punnett square predicts that

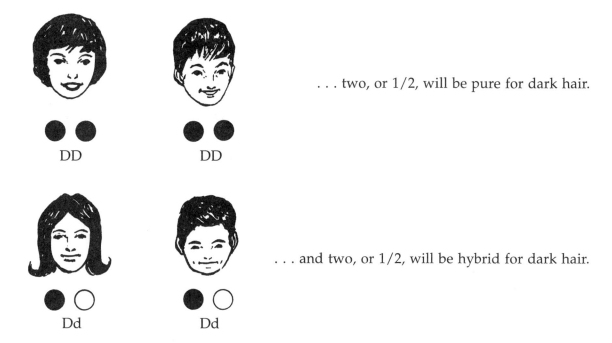

. . . two, or 1/2, will be pure for dark hair.

. . . and two, or 1/2, will be hybrid for dark hair.

Remember, there are only two possibilities: <u>pure</u> dark and <u>hybrid</u> dark. And you cannot tell by looking at the children which ones are pure and which ones are hybrid for dark hair.

Which gene combinations will turn up in a child? It's a matter of chance.

PREDICTING HEREDITY IN PEA PLANTS

When Mendel did his experiments with pea plants he found that some peas had a smooth covering. Others were wrinkled.

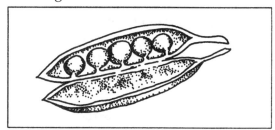

Figure A *Smooth peas are dominant (S).*

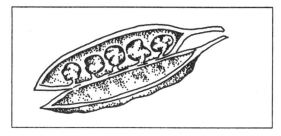

Figure B *Wrinkled peas are recessive (s).*

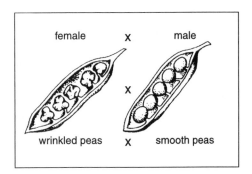

Figure C

Let's see what happens when a pure smooth pea plant is crossed with a pure wrinkled pea plant.

1. The dominant smooth genes come from the

   _____male_____ .
      male, female

2. The recessive wrinkled genes come from the

   _____female_____ .
      male, female

male gametes

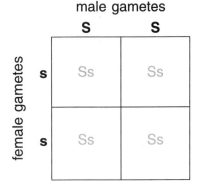

Figure D

3. Now fill in the Punnett square for Figure D.

4. What kind of covering will all the offspring

   peas have? _____smooth_____
                smooth, wrinkled

5. All the offspring are _____hybrids_____ .
                        pure, hybrids

male gametes

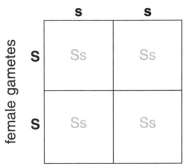

Figure E

6. Would there be any difference if the dominant (S) genes were from the female and the recessive (s) genes were from the male? Test it in Figure E.

   Answer: There _____would not_____ be a
   difference.   would, would not

Now try a cross between two hybrids, Ss x Ss. Fill in Figure F.

7. How many offspring will be smooth?

   _____3 out of 4_____

8. How many will be wrinkled? _____1 out of 4_____

9. How many offspring will be pure smooth?

   _____1_____

10. How many will be hybrid smooth? _____2_____

Figure F

Figure G

Gary and Tina are married. They are planning a family. What will their children look like? Try some more Punnett squares to find out.

Gary is hybrid for curly hair (Cc). Tina is pure for straight hair (cc).

> C = dominant curly
> c = recessive straight

Gary is hybrid for dark hair (Dd). Tina is pure for blonde hair (dd).

> D = dominant dark
> d = recessive blonde

Both Gary and Tina are hybrid for brown eyes (Bb).

> B = dominant brown
> b = recessive blue

Complete the Punnett square for each trait. Then answer the questions.

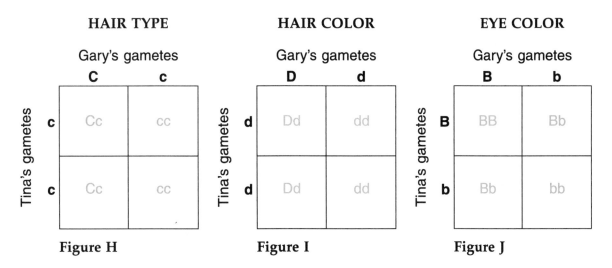

HAIR TYPE	HAIR COLOR	EYE COLOR

Figure H   **Figure I**   **Figure J**

1. How many offspring will have curly hair? _____2 out of 4_____

2. How many offspring will have straight hair? _____2 out of 4_____

3. How many offspring will be pure for curly hair? _____none_____

4. How many offspring will be pure for straight hair? _____2 out of 4_____

5. How many will be hybrid for curly hair? _____ 2 out of 4 _____

6. How many offspring will have dark hair? _____ 2 out of 4 _____

7. How many offspring will have blonde hair? _____ 2 out of 4 _____

8. How many offspring will be pure for dark hair? _____ 0 _____

9. How many will be pure for blonde hair? _____ 2 out of 4 _____

10. How many will be hybrid for dark hair? _____ 2 out of 4 _____

11. How many offspring will have brown eyes? _____ 3 out of 4 _____

12. How many offspring will have blue eyes? _____ 1 out of 4 _____

13. How many will be pure for brown eyes? _____ 1 out of 4 _____

14. How many will be pure for blue eyes? _____ 1 out of 4 _____

15. How many will be hybrid for brown eyes? _____ 2 out of 4 _____

MATCHING

*Match each term in Column A with its description in Column B. Write the correct **letter in the** space provided.*

	Column A		**Column B**
e	1. Punnett square	a)	represented by a lower **case letter**
c	2. dominant gene	b)	male gametes
d	3. egg cells	c)	represented by a capital **letter**
b	4. sperm cells	d)	female gametes
a	5. recessive gene	e)	used to show gene **combinations**

What is incomplete dominance?

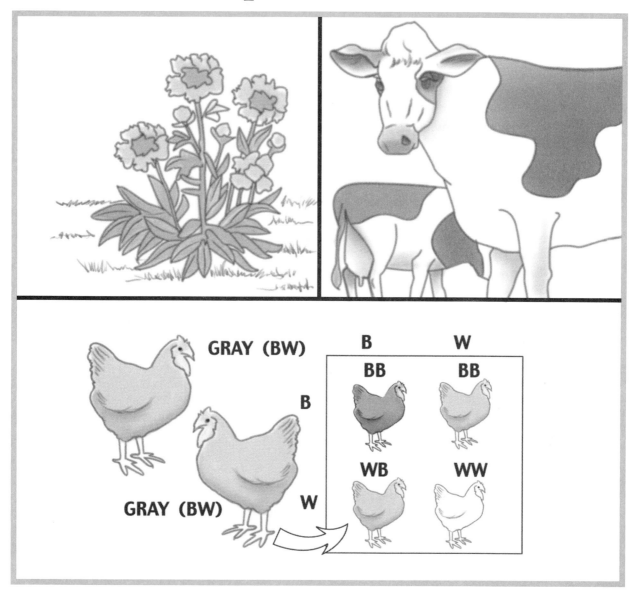

KEY TERMS

incomplete dominance: blending of traits carried by two or more different genes

blending: combination of genes in which a mixture of both traits shows

LESSON 5 | What is incomplete dominance?

In most games, there is a stronger team and a weaker team. Usually, the stronger team wins. Some games end in ties. This shows that the teams were equally matched.

Heredity is sometimes like this. Most traits have a stronger, dominant gene, and a weaker, recessive gene. The dominant gene usually "wins." The dominant trait shows up in the offspring. The recessive trait stays "hidden."

Not all genes, however, are completely dominant or completely recessive. The genes of certain traits are equally strong. Neither trait is dominant. We say there is **incomplete dominance**. In cases of incomplete dominance, genes combine and a mixture of both traits shows up. This kind of gene combination is called **blending**.

Three good examples of incomplete dominance are found in the colors of four-o'clock flowers, shorthorn cattle, and Andalusian [an-duh-LEW-zhun] fowl.

Four-o'clock flowers Four-o'clock flowers are usually red or white. Red and white are equally strong traits. Neither color is dominant. When a pure red (RR) crosses with a pure white (WW), the colors blend. The offspring have pink flowers (RW).

Shorthorn cattle In cattle, if one parent is pure red (RR) and the other parent is pure white (WW), the offspring will be pink—a blend of red and white (RW). The "blended" calf is called a <u>roan</u> calf.

Andalusian fowl Some of these chickens have genes for black feathers. Others have genes for white feathers. Neither of these genes is dominant. The offspring of pure black and pure white Andalusians are gray. Gray is a blend of black and white.

Many genes in humans also show incomplete dominance. They include genes for most hair and eye colors.

Figures A, B, and C show three examples of incomplete dominance.

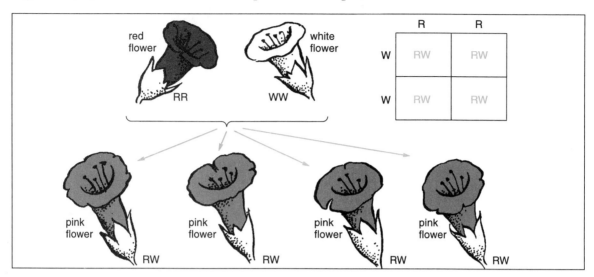

Figure A *The crossing of pure red (RR) and pure white (WW) four-o'clock flowers.*

1. Fill in the Punnett square in Figure A.

_____c_____ 2. The offspring of crossed pure red and pure white four-o'clock flowers are
   a) only red.
   b) only white.
   c) only pink.
   d) both red and white.

_____c_____ 3. In four-o'clock flowers,
   a) red is dominant over white.
   b) white is dominant over red.
   c) neither red nor white is dominant.
   d) pink is dominant over red.

4. Pink is a blend of which two colors? _____red_____ and

_____white_____

_____c_____ 5. Blended four-o'clock flowers have
   a) only genes for the color white.
   b) only genes for the color red.
   c) genes for both red and white.
   d) only pink genes.

6. Blended four-o'clock flowers are _____hybrids_____ .
   pure, hybrids

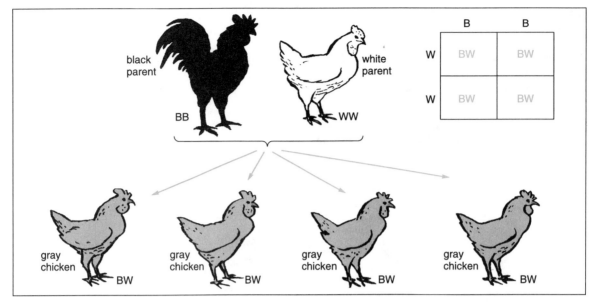

Figure B *The crossing of pure black (BB) and pure white (WW) Andalusian chickens.*

7. Complete the Punnett square in Figure B.

_____c_____ 8. The offspring of crossed pure black and pure white Andalusian chickens are

    a) only white.   **b)** only black.

    c) a blend of black and white.   **d)** black and white.

_____c_____ 9. In Andalusian chickens,

    a) black is dominant over white.   **b)** white is dominant over black.

    c) neither black nor white is dominant.   **d)** both black and white are dominant.

10. What color are the offspring of black and white chickens?

    _____gray_____

11. Gray is a blend of which two colors? _____white_____ and

    _____black_____

_____c_____ 12. Blended Andalusian fowl have

    a) only genes for the color black.   **b)** only genes for the color white.

    c) genes for both black and white.   **d)** only gray genes.

13. Blended Andalusian fowl are _____hybrids_____ .
                                    pure, hybrids

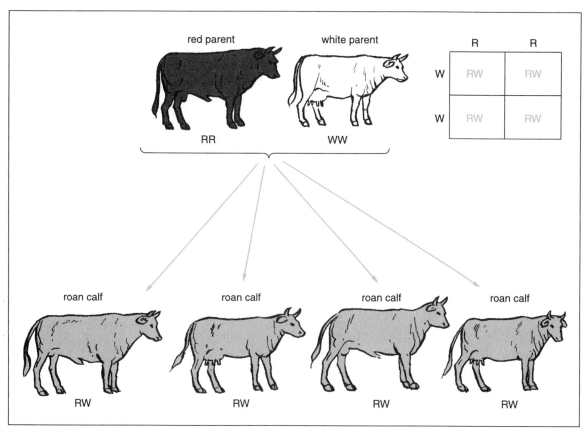

Figure C *The crossing of pure red (RR) and pure white (WW) shorthorn cattle.*

14. Complete the Punnett square in Figure C.

_____c_____ 15. The offspring of crossed pure red and pure white shorthorn cattle are

    a) only red.   **b)** only white.

    c) a blend of red and white.   **d)** red and white.

_____c_____ 16. In shorthorn cattle,

    a) red is dominant over white.   **b)** white is dominant over red.

    c) there is incomplete dominance of red and white colors.   **d)** white is recessive to red.

17. What are red and white blended cattle called? _____roan cattle_____

_____c_____ 18. Roans have

    a) only genes for the color white.   **b)** only genes for the color red.

    c) genes for both white and red.   **d)** only dominant genes.

19. Roans are _____hybrids_____ .
                    pure, hybrids

31

Complete each statement using a term or terms from the list below. Write your answers in the spaces provided. Some words may be used more than once.

blending	red	pure
roan calf	eye	hybrid
recessive	pink four-o'clock flower	strong
white	incomplete dominance	
dominant	hair	

1. A "hidden trait" is called a ____recessive____ trait.

2. Not all genes are completely recessive nor completely ____dominant____. Some are equally ____strong____.

3. An individual that has only dominant or recessive genes for a trait is ____pure____ for that trait.

4. An individual that has both dominant and recessive genes for a trait is ____hybrid____ for that trait.

5. A condition where the genes for a given trait are equally strong is called ____incomplete dominance____.

6. A combination of genes in which a mixture of both traits shows up is called ____blending____.

7. Two examples of offspring of incomplete dominance are ____roan calf____ and the ____pink four-o'clock flower____.

8. In four-o'clock flowers and roan cattle, neither the color ____red____ nor the color ____white____ is dominant.

9. Incomplete dominance produces offspring with ____hybrid____ genes for the given trait.

10. Examples of incomplete dominance in humans are found in ____hair____ and ____eye____ color.

MATCHING

Match each term in Column A with its description in Column B. Write the correct letter in the space provided.

	Column A		Column B
__d__	1. RR	a)	recessive pure
__e__	2. Rr	b)	control heredity
__a__	3. rr	c)	blends traits
__c__	4. incomplete dominance	d)	dominant pure
__b__	5. genes	e)	hybrid

Complete the Punnett square for feather color in chickens.

B = black feathers

W = white feathers

BW = gray feathers

	W	**W**
B	BW	BW
B	BW	BW

1. Are the parents in this cross pure or hybrids? _____ pure
2. What color are the parents? _____ One parent is black and the other is white.
3. Will all of the offspring produced by this cross be hybrids? _____ yes
4. What colors will the offspring be? _____ gray
5. Why is neither gene in this cross represented by a lower case letter? _____ In cases of incomplete dominance both genes are represented by capital letters.

WORD SCRAMBLE

Below are several scrambled words you have used in this lesson. Unscramble the words and write your answers in the spaces provided.

1. TOMANDIN _____ DOMINANT
2. SEERVICES _____ RECESSIVE
3. DLEBN _____ BLEND
4. RATTI _____ TRAIT
5. DHSYIRB _____ HYBRIDS

SCIENCE EXTRA

Condor Care

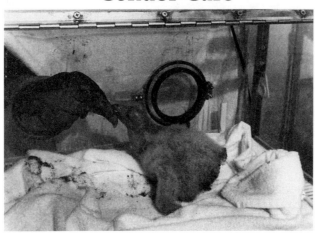

They are described as ugly by some and beautiful by others. But despite their looks, California Condors have fought off the threat of extinction.

Of course, the condors have had some help. The first sign that the condors were in danger came in 1939. Scientists noticed that the number of California Condors was going down. Too many condors were dying for the wrong reasons. Some birds were being shot. Others were being poisoned by chemicals used on crops. Still others were dying because their habitat was being destroyed.

By 1967, only about 50–60 birds were left in the wild. This placed the condor on the first Endangered Species List. Numbers continued to go down, to 25–35 birds by 1980. Then the birds reached their lowest number, 9, in 1985.

The San Diego and Los Angeles Zoos get most of the credit for saving the condors. Both zoos were quick to set up special breeding and foster-parenting programs. In 1981, a special breeding area was built at San Diego Zoo's Wild Animal Park. It was called the "Condorminium." It was

here that chicks were successfully fed by lifelike hand puppets. These are not ordinary puppets. They are made to look like the neck and head of a mother condor.

A new chick is never allowed to see humans. This is because the chicks will "imprint" (form a strong attachment) on whatever they see early in life. A chick must seek the company of condors, not humans, to live.

Puppet feeding begins on day three of the chick's life. The keeper uses the hand puppet to feed chopped food to the chick. The puppet is even used to teach the chick how to eat. For the first 28 days, only the condor puppet feeds the chick. The puppet also interacts with and parents the chick. Then the chick is transferred to an outdoor enclosure where it can grow and develop.

The "Condorminium" has been helping to save condors for almost 20 years. Since the program began, California Condors have been set free in California and the Grand Canyon. Today, there are about 120 condors living both in the wild and in captivity.

How is sex determined?

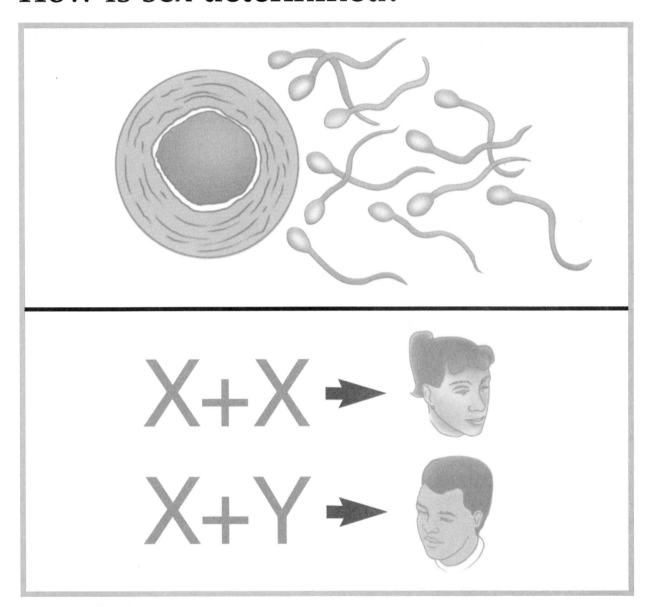

KEY TERM

sex chromosomes: X and Y chromosomes

LESSON 6 | How is sex determined?

Will the baby be a boy or a girl? Every expectant parent asks this question. Well, the possibilities are even—50 percent for a boy, 50 percent for a girl. It depends entirely upon chance.

A single family may have more girls than boys, or more boys than girls. The population on the whole is just about even—half male and half female. Let us find out why.

Look at the 23 pairs of human chromosomes below.

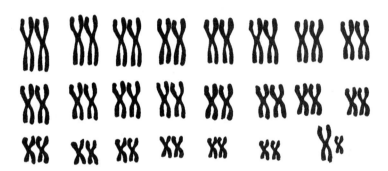

Chromosomes from a body cell of a human male

Notice that each chromosome in a pair is the same size and shape—except for the last pair. In a male, the chromosomes in the last pair are different. The larger is called the X chromosome. The smaller is the Y chromosome. X and Y chromosomes are the **sex chromosomes**. They determine the sex of most organisms.

- A male cell has one X chromosome and one Y chromosome (XY).

- A female cell has two X chromosomes (XX).

What determines the sex of an offspring? The next page tells the story. We will use the fruit fly as an example.

A fruit fly body cell has eight chromosomes (four pairs). Two of these (one pair) are sex chromosomes. Special body cells produce gametes, or sex cells, by the process of <u>meiosis</u>. Meiosis is a special kind of cell division.

FRUIT FLY

Fruit fly nucleus
before meiosis
= 8 chromosomes

egg or sperm nuclei
in sex cells
= 4 chromosomes each

Figure A

In a male, meiosis produces four sperm cells from one body cell. During meiosis, each sperm cell receives only one sex chromosome from a pair. In a female, meiosis produces one usable egg cell and three unusable cells from one body cell. During meiosis, each egg cell receives one sex chromosome from a pair.

HOW EGG AND SPERM CELLS ARE PRODUCED IN FRUIT FLIES

Gamete chromosomes are not paired. They are single chromosomes. A gamete, then, has half the number of chromosomes of a body cell. Count them in Figure A.

Look at the sex chromosome of each gamete.

- An egg has an X chromosome only.

- A sperm may have an X or a Y chromosome; 50 percent have an X chromosome; 50 percent have a Y chromosome.

Now here is where chance comes in.

- If an X sperm fertilizes an egg (X + X), the offspring will be a female (XX).

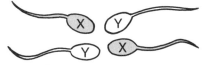

- If a Y sperm fertilizes an egg (Y + X), the offspring will be a male (XY).

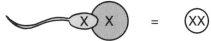

As a result, offspring inherit their sex from their father. Since half of sperm carry the X chromosome and half carry the Y chromosome, over a large number of births, half will be female and half will be male.

Male or female? The chances are 50-50. See for yourself in Figure B.

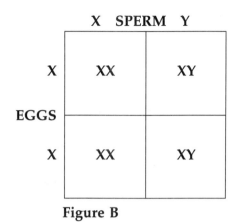

X   SPERM   Y

Figure B

XX= female — 50 percent
XY = male — 50 percent

Sperm have tails and are able to swim. Millions of sperm may swim toward the same egg, but only one can fertilize it.

Will it be an X sperm or a Y sperm? It depends upon chance!

1. If an X sperm fertilizes the egg, the

   offspring will be a _____female_____ .
   <u>male, female</u>

2. If a Y sperm fertilizes the egg, the

   offspring will be a _____male_____ .
   <u>male, female</u>

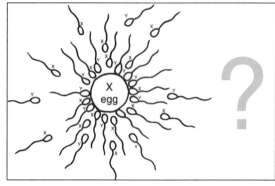

Figure C

Study the human sex chromosomes shown. Then, in the space provided, identify whether a child with these sex chromosomes is a male child or a female child.

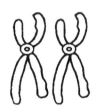

Figure D

3. _____female_____

Figure E

4. _____male_____

5. Explain your answers. ___Figure D shows two X chromosomes; Figure E___

   __shows one X chromosome and one Y chromosome.__

   _____

38

Complete each statement using a term or terms from the list below. Write your answers in the spaces provided. Some words may be used more than once.

chance	X	gametes
female	half	50-50
Y	male	

1. There are two kinds of sex chromosomes. They are called _____X_____ and

   _____Y_____ .

2. A female body cell has only _____X_____ sex chromosomes.

3. A male body cell has both _____X_____ and _____Y_____ sex chromosomes.

4. Meiosis produces _____gametes_____ .

5. A gamete has _____half_____ the number of chromosomes found in a body cell.

6. Eggs have only _____X_____ sex chromosomes.

7. Sperm have either _____X_____ or _____Y_____ sex chromosomes.

8. Which will fertilize an egg, an X sperm or a Y sperm? It depends entirely upon

   _____chance_____ . The odds are _____50-50_____ .

9. The fertilization of an egg by an X sperm produces a _____female_____ offspring.

10. The fertilization of an egg by a Y sperm produces a _____male_____ offspring.

MATCHING

Match each term in Column A with its description in Column B. Write the correct letter in the space provided.

	Column A		Column B
__b__	1. X and Y chromosomes	a)	XY
__a__	2. male	b)	sex chromosomes
__d__	3. meiosis	c)	XX
__e__	4. sperm	d)	special cell division
__c__	5. female	e)	male gamete

TRUE OR FALSE

In the space provided, write "true" if the sentence is true. Write "false" if the sentence is false.

True   **1.** A body cell has paired chromosomes.

False   **2.** A gamete has paired chromosomes.

False   **3.** An egg has only a Y sex chromosome.

True   **4.** A sperm can have either an X or a Y sex chromosome.

False   **5.** An X chromosome looks the same as a Y chromosome.

False   **6.** Many sperm fertilize one egg.

True   **7.** Fertilization by an X sperm produces a female.

True   **8.** Fertilization by a Y sperm produces a male.

True   **9.** Humans have 23 pairs of chromosomes.

True   **10.** About the same number of male and female organisms are born.

FLIP A COIN

Flip a coin 100 times. Count how many times it lands on heads, and how many times it lands on tails.

1. How many times did it land on heads? _____ Answers will vary.

2. How many times did it land on tails? _____ Answers will vary.

3. Is it about 50-50? _____ yes

4. Is flipping a coin a chance event? _____ yes

5. What other chance event did you learn about in this lesson? _____ whether an X or

Y sperm fertilizes an egg

40

How does the environment affect traits?

LESSON 7 | How does the environment affect traits?

What makes you the way you are? Genes? Of course! Genes control many of your traits. Genes do not work alone. The environment also affects the traits of living things.

The environment is made up of all the living and nonliving things that surround an organism. Air, water, temperature, and food are all parts of the environment. The right combination of these things is needed for an organism to develop properly. When the environment is not right for an organism, certain traits may not develop at all.

Green plants, for example, need an environment of proper sunlight, temperature, water, and minerals to grow well. In a poor environment, they grow small, weak, and pale. The size of a plant is a trait. The gene for this trait is not affected by the environment, but the development of the trait is.

Suppose a person has the genes for growing tall. A poor diet may prevent full growth.

Diet is just one part of your environment. There are many other things that make up your environment. Can you name some that have not been listed?

We cannot change our genes. But, we can change some parts of the environment in which we live. Eating, resting, and exercising properly help the traits we are born with develop to their fullest.

Traits need a proper environment to develop properly.

Figure A

Figure B

1. In which figure do you see a plant that grew in a good environment? _____A_____

2. How can you tell? _____It looks healthy._____

_____

3. In which figure do you see a plant that grew in a poor environment? _____B_____

4. How can you tell? _____It looks like it is dying. Its stems and leaves are drooping._____

_____

These two rats were born in the same litter. One ate a poor diet; the other ate a diet rich in nutrients.

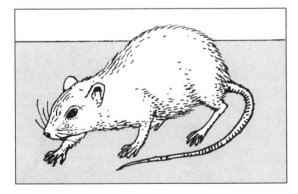

Figure C

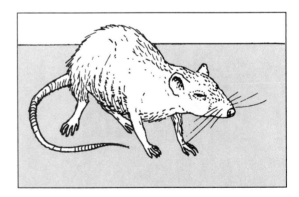

Figure D

5. Is diet part of a living thing's environment? _____yes_____

6. Which figure shows the rat that ate a poor diet? _____D_____

7. How do you know? _____It looks sickly._____

8. Which figure shows the rat that ate a good diet? _____C_____

43

9. How do you know? _It looks more healthy and alert._

10. In your own words, what does <u>environment</u> mean? _Answers will vary, but should include everything that surrounds an organism._

ARE ALL TRAITS INHERITED?

Many traits are not inherited. These traits are called acquired traits.

Here are some examples of acquired traits.

Figure E *Some people develop strong muscles by exercising.*

Figure F *Some people learn to speak many languages.*

Figure G *The tails of some dogs are cut off at birth.*

Figure H *Trees near the tops of mountains do not grow tall partly because of low temperatures.*

WHAT DO YOU THINK?

Can acquired traits be passed on to an offspring? In other words, can acquired traits be inherited? Answer this question yourself.

1. If parents develop strong muscles by exercising, will their children inherit strong

   muscles? _____ no _____
              yes, no

2. Suppose you have learned to speak a new language? Will your children be born

   with this ability? _____ no _____
                       yes, no

3. Suppose the dog in Figure G has puppies. Will they be born without tails? ___ no ___

4. Conclusion: Acquired traits _____ are not _____ inherited.
                                     are, are not

5. What kind of traits do you think are inherited? _____ traits controlled by genes _____

FILL IN THE BLANK

Complete each statement using a term or terms from the list below. Write your answers in the spaces provided. Some words may be used more than once.

living	traits	inherited
acquired	nonliving	environment
genes	good	

1. The characteristics of an individual are called _____ traits _____ .

2. Inherited traits are passed on to offspring by _____ genes _____ .

3. Some traits develop properly only in a proper _____ environment _____ .

4. The environment includes all the _____ living _____ and _____ nonliving _____ things surrounding an organism.

5. Traits that are not inherited are called _____ acquired _____ traits.

6. Acquired traits are not _____ inherited _____ .

7. Traits develop best in a _____ good _____ environment.

8. We cannot change our _____ genes _____ .

9. A trait that is not carried by genes is an _____ acquired _____ trait.

10. Air, water, food, and temperature are all parts of the _____ environment _____ .

MATCHING

Match each term in Column A with its description in Column B. Write the correct letter in the space provided.

	Column A	**Column B**
_a_	**1.** acquired traits	**a)** not inherited
_c_	**2.** environment	**b)** genes
_d_	**3.** proper environment	**c)** everything surrounding an individual
_b_	**4.** carriers of inherited traits	**d)** best for developing traits
_e_	**5.** trait	**e)** any characteristic of a living thing

TRUE OR FALSE

In the space provided, write "true" if the sentence is true. Write "false" if the sentence is false.

False   **1.** Every trait is carried by genes.

False   **2.** Every trait is inherited.

True   **3.** A trait that is not carried by genes is an acquired trait.

False   **4.** Offspring inherit the acquired traits of their parents.

True   **5.** Environment affects how traits develop.

False   **6.** Traits develop best in a poor environment.

True   **7.** We can control part of our environment.

False   **8.** You can inherit genes for facts you learn.

True   **9.** Exercise can make strong muscles even stronger.

False   **10.** Extra-strong muscles developed by exercise can be inherited.

What are some methods of plant and animal breeding?

KEY TERMS

controlled breeding: mating organisms to produce offspring with certain traits

mass selection: crossing organisms with desirable traits

inbreeding: mating closely related organisms

hybridization: mating two different kinds of organisms

LESSON 8 | What are some methods of plant and animal breeding?

You have probably eaten corn on the cob. Have you ever eaten an ear of corn that is only as long as your thumb? Perhaps you have, but we usually think of corn as a large-sized vegetable.

Tiny ears of corn were all that existed thousands of years ago. However, the Indians in South and Central America changed that. They noticed that, in nature, some ears of corn were larger than others. So they crossed the plants with the largest ears. They found that the offspring were likely to have large ears, too. For many years, the Indians selected only the seeds from the largest ears to reproduce. As a result, the size of corn ears greatly increased. A new variety of plant had developed.

The mating of animals and plants to produce certain desirable traits is called **controlled breeding**. People bred animals and plants long before they knew about chromosomes and genes. Now we know much more about genetics. We use this information to produce organisms that are useful to people.

Breeding helps supply the hungry world with food. Plants can be bred to produce larger and better crops. Animals can be bred to produce more and better meat, milk, and wool.

Breeding also serves special needs. For example, breeding has given us giant flowers with unusual colors. We now also have faster racehorses. Even dogs are bred for special jobs such as protection and to help guide the blind.

Sometimes scientists breed plants and animals in laboratories. There, chromosomes can be studied under powerful microscopes. These experiments have led to many discoveries. As a result of these discoveries, scientists can now detect and control certain diseases. Our knowledge of how traits are passed on has certainly increased since Mendel's time.

Figures A through D show methods of controlled breeding. Study each figure and answer the questions.

Figure A *Mass selection*

Figure B *Inbreeding*

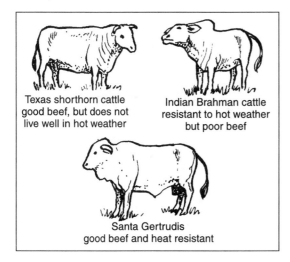

Figure C *Cross-breeding*

To grow corn with large ears, the Indians of South and Central America used a method of controlled breeding. They crossed plants with good traits and then collected and planted the seeds for many plant generations.

Today we call this method **mass selection**.

1. What is the purpose of mass selection?

   To develop a new kind of plant with

   certain good traits.

Another breeding method is **inbreeding**. Inbreeding is the mating of closely related organisms.

Inbreeding is used to keep certain kinds of animals pure. For example, thoroughbred racehorses are bred for speed.

2. Inbreeding produces organisms with

   very _____similar_____ genes.

   similar, different

Sometimes, related but different breeds of animals or plants are mated. This combines desired traits.

3. What kind of climate do you think the Indian Brahman came from?

   _____hot_____

4. **a)** If you were a Texas cattle rancher, which breed would you want?

   Santa Gertrudis

   b) Why? It produces good beef and

   is heat resistant.

49

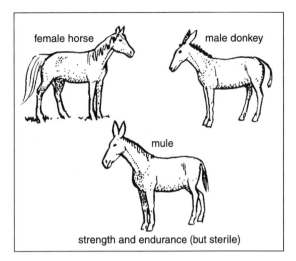

Figure D *Hybridization*

Sometimes two different species of plants or animals are mated. This is called **hybridization** [hy-brid-ih-ZAY-shun]. This can also combine desirable traits. However, the offspring are usually sterile. They are unable to reproduce.

5. Are mules able to reproduce?

_____ No _____

Figure E

The controlled breeding of animals and plants helps increase food production.

6. Why is greater food production important to the world's population?

Greater food production

is important for people

who do not have

enough to eat.

Breeding improves quality and gives us plants and animals with special traits.

Figure F

7. Do you think that all the food you eat is the same as the food people ate a hundred years ago? _____ No _____

50

Decide which of the statements refers to mass selection (M), inbreeding (I), or hybridization (H). In the spaces provided, write the correct letter for each statement.

_____I_____   **1.** Crossing of closely related organisms

_____M_____   **2.** Planting seeds that show desired traits

_____H_____   **3.** Organisms used are genetically different

_____M_____   **4.** Crossing of plants with desired traits

_____I_____   **5.** Self-pollination in plants

_____H_____   **6.** Offspring in a male lion and female tiger

_____H_____   **7.** Crossing wheat and rye plants

_____M_____   **8.** New varieties of what were bred to produce more protein

_____I_____   **9.** Have genes very similar to their parents

_____I_____   **10.** Purebred dog

NOW TRY THESE

Read the examples. Complete the table by writing the letters of the examples in the correct columns.

Examples

  a. Farmer wants sweet corn seeds that will produce tall plants with a high yield.

  b. Seed producer wants to develop corn that will resist drought.

  c. Dog breeder wants purebred dogs.

  d. Offspring of a male donkey and a female horse.

  e. Florist wants roses with large petals.

Controlled Breeding

	Method	Example
1.	Mass selection	a, b, e
2.	Inbreeding	c
3.	Hybridization	d

In the space provided, write "true" if the sentence is true. Write "false" if the sentence is false.

False	**1.**	Every farm acre grows the same size crop.
False	**2.**	Every cow gives the same amount of milk.
True	**3.**	Breeding is controlled reproduction.
True	**4.**	Breeding helps increase our food supply.
True	**5.**	In breeding, the best animals are selected for reproduction.
True	**6.**	Breeders select organisms with the best traits for breeding.
False	**7.**	All wheat plants are the same.
True	**8.**	Breeding can develop animals that are more resistant to heat.
True	**9.**	Mules are sterile.
False	**10.**	Inbreeding produces organisms with very different genes.

MATCHING

Match each term in Column A with its description in Column B. Write the correct letter in the space provided.

Column A

c	**1.**	controlled breeding
a	**2.**	inbreeding
e	**3.**	mass selection
b	**4.**	cross-breeding
d	**5.**	hybridization

Column B

a) mating only closely related organisms

b) mating related, but different breeds

c) methods used to produce organisms with desirable traits

d) mating different species of animals

e) used to develop new plant varieties

REACHING OUT

Inbreeding produces organisms that are very similar genetically. They have few genetic differences. Why do you think this is a problem to a species? __Answers will vary.__

Accept all logical responses.

Genetics

Lesson

What is genetic engineering?

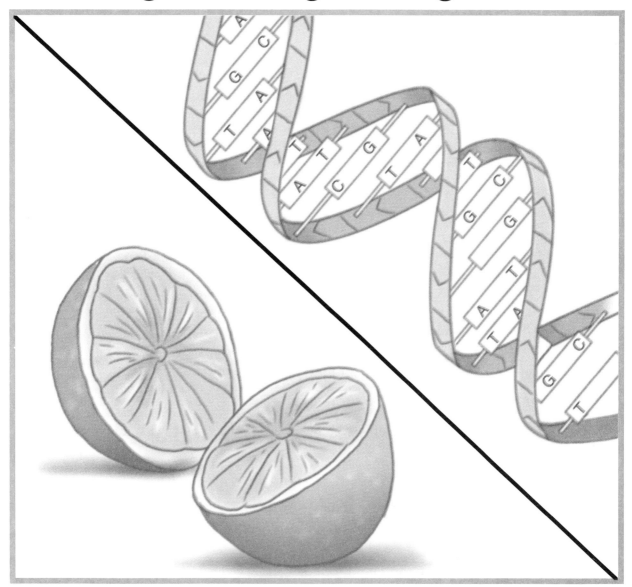

KEY TERMS

genetic engineering: methods used to produce new forms of DNA

gene splicing: moving a section of DNA from the genes of one organism to the genes of another organism

cloning: production of organisms with identical genes

LESSON 9 | What is genetic engineering?

Have you ever seen a supermouse? Supermouse is not a character in a cartoon. It is the nickname given to a mouse produced by researchers. Supermouse is twice the size of a normal mouse. It is the result of a new technology called **genetic engineering** [juh-NET-ik en-juh-NEER-ing]. In genetic engineering, scientists work with individual genes.

In Lesson 2, you learned that genes are made up of a complex substance called <u>DNA</u>. Genetic engineering is a process by which new forms of DNA are made.

One method of genetic engineering is called **gene splicing** [SPLYS-ing]. Gene splicing is the process by which pieces of DNA from the genes of one organism are transferred to another organism. This process does not involve cutting apart genes with tiny instruments. Instead, it makes use of chemicals that break or combine pieces of DNA in precise ways.

Gene splicing takes place in three steps. Look at Figures A through D on the next page as you read about these steps.

1. A DNA chain is opened up.

2. New genes from another organism are added, or spliced, into the DNA.

3. The DNA chain is closed.

Once the genes are transferred, they become part of the receiving organism's genes. As a result, the trait carried by the genes is passed on to future generations.

Genetic engineering can help produce useful plants and animals. It can also help solve many health problems. However, this process might also produce new microorganisms that cause diseases that we cannot control. To prevent such disasters, scientists and lawmakers are working together to establish rules and safeguards.

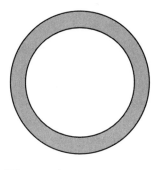

A ring of DNA.

Figure A

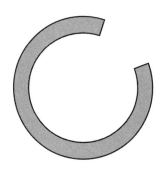

The DNA chain is opened up.

Figure B

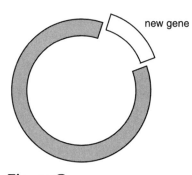

new gene

A new strand of DNA, or gene, is added into the ring of DNA.

Figure C

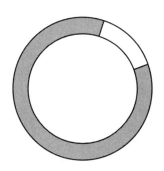

The DNA chain is closed.

Figure D

Through genetic engineering, scientists have been able to splice human genes into the DNA of bacteria. The bacteria can then produce substances that otherwise could only be made by the human body. Here are some substances produced by bacteria through genetic engineering.

INSULIN [IN-suh-lin]   Insulin is needed by people with diabetes [dy-uh-BEET-is]. Insulin controls the level of sugar in your blood.

HUMAN GROWTH HORMONE   Human growth hormone controls growth. It is given to children who do not make enough of their own growth hormone. This helps the children grow properly.

INTERFERON [in-tur-FEER-ahn]   Interferon helps your body fight disease. It is used by scientists in cancer research.

Scientists also hope that someday genetic engineering can be used to correct some genetic disorders. They may be able to add normal genes to cells that have abnormal genes or are missing a gene completely.

TRUE OR FALSE

In the space provided, write "true" if the sentence is true. Write "false" if the sentence is false.

True	**1.** Once genes are transferred during gene splicing, they become part of the receiving organism's genes.
False	**2.** Genetic engineering is an old technology.
False	**3.** Insulin controls growth.
True	**4.** New forms of DNA are made in genetic engineering.
True	**5.** Genes are made up of DNA.
False	**6.** Interferon controls the level of sugar in the blood.
True	**7.** A chain of DNA is opened up during gene splicing.
True	**8.** Some children do not produce enough human growth hormone.
False	**9.** Closing a chain of DNA is the first step of gene splicing.
True	**10.** Bacteria are used in genetic engineering.

Scientists have other ways to change the genes of organisms besides genetic engineering. Have you ever eaten a seedless orange? Seedless oranges are produced by **cloning**. Cloning is the production of organisms with identical genes. Clones are produced by asexual reproduction.

The first seedless orange tree was the result of a <u>mutation</u> [myoo-TAY-shun]. A mutation is a sudden change in genes. A mutation is an accident. It can cause new inherited traits.

Most mutations occur in nature.

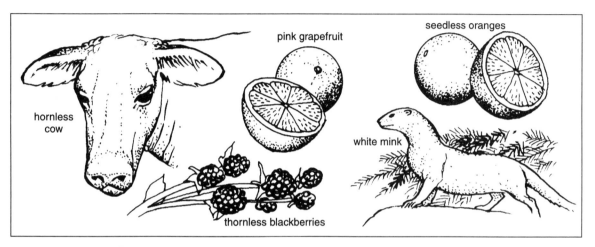

Figure E *Some familiar mutations*

Figure F *The Mediterranean fruit fly damages fruit.*

Scientists can also cause mutations in the laboratory with radiation.

Some male insect pests, like the Mediterranean fruit fly, are given radiation. The radiation causes many changes in the genes. The genes become damaged. With damaged genes, the male flies are sterile.

When the flies mate, no offspring are produced. The insect population is reduced. This is a way of reducing insect pests without harmful chemicals.

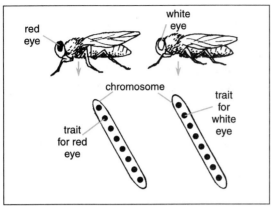

Figure G *White eyes is a mutant trait in fruit flies.*

Mutations also help scientists make a "map" of the chromosomes of the fruit fly.

A chromosome map shows which part of the chromosome controls which trait.

Match each term in Column A with its description in Column B. Write the correct letter in the space provided.

Column A

___e___   **1.** mutation

___d___   **2.** chromosome map

___b___   **3.** radiation

___a___   **4.** cloning

___c___   **5.** seedless orange

Column B

a) production of organisms with identical genes

b) can damage genes

c) result of mutation

d) shows which part of the chromosome controls a trait

e) sudden change in genes

REACHING OUT

Pieces of DNA that contain DNA from a different organism are called <u>recombinant DNA</u>. Why do you think this is a good name?

Answers will vary. Accept all logical responses.

What is natural selection?

KEY TERMS

fossils: remains of organisms that lived in the past

extinct: organism that no longer exists on earth

evolution: process by which organisms change over time

natural selection: survival of organisms with favorable traits

LESSON 10 | What is natural selection?

Have you ever visited a natural history museum? If you have, you probably saw some **fossils**. Fossils are the remains of organisms that lived in the past.

Until the 19th century, most scientists believed that organisms had not changed since the time they first appeared on the earth. However, by the late 1700s, scientists had found and studied many fossils. Fossils show interesting things about living things.

Fossils show that organisms have changed. They show that the earliest living things on the earth were simple organisms. In the billions of years that passed, living things became more complex.

Fossils show that many species, or kinds of organisms, died out. These organisms are **extinct**.

Most scientists believe that new species develop from old species as a result of gradual change or **evolution** [ev-uh-LOO-shun]. Evolution is the process by which organisms change over time.

How and why have living things changed? Different theories of evolution have been given over the years. However, over 100 years ago, an English biologist name Charles Darwin suggested a theory of evolution. Darwin's theory is accepted by most scientists today.

According to Darwin's theory:

1. **OVERPRODUCTION**   Organisms produce more offspring than the environment can support. There is not enough food or living space for all of the offspring.

Figure A

2. **COMPETITION**   Overproduction leads to a struggle. All the organisms compete for food, water, and the other necessities of life. Only those organisms that are well suited to their surroundings survive and reproduce. The rest die.

Figure B

3. **VARIATIONS** Organisms of the same species are very similar. But they do have individual differences among traits, or variations. These differences are important in the "struggle for survival." For example, extra speed can mean the difference between life and death. A fast wildebeest may escape an attacking lion. A slower neighbor may become the lion's next meal.

Figure C

4. **SURVIVAL OF THE FIT** Organisms with traits that make them well adapted, or suited to the environment, survive and reproduce. Darwin used the term **natural selection** to describe the survival of organisms with favorable traits. They, in turn, pass their favorable traits to their offspring. The offspring are then more likely to survive. As the process of natural selection goes on over many generations, species change. These changes can result in the appearance of a new species. Evolution by natural selection occurs.

Figure D

Study the diagrams. Then answer the questions.

1. According to Darwin, did all ancient giraffes have long necks?

   _____no_____

2. _____Long_____ -necked giraffes
   <u>Long, Short</u>
   were better able to reach food far off the ground.

Figure E

3. _____Long_____ -necked giraffes
   <u>Long, Short</u>
   were better suited to the environment.

4. The _____short_____ -necked
   <u>long, short</u>
   giraffes died out.

Figure F

5. Which giraffes survived and reproduced?

   The _____long_____ -necked
   <u>long, short</u>
   giraffes.

6. What important adaptation did the surviving giraffes pass on to their offspring?

   _____long necks_____

7. Describe the necks of all giraffes living today (HINT: One word will do).

   _____long_____

Figure G

Complete each statement using a term or terms from the list below. Write your answers in the spaces provided.

changing   favorable   organisms
Charles Darwin adapted   competition
variations   different   limited number
extinct   reproduce

1. An organism that is suited to its environment is said to be _____adapted_____ to its surroundings.

2. Earth is always _____changing_____ .

3. As the earth changes, the _____organisms_____ that live on it also change.

4. A species that does not change as its environment changes may become

   _____extinct_____ .

5. The scientist who developed an important theory of evolution was

   _____Charles Darwin_____ .

ACCORDING TO DARWIN:

6. A favorable environment can support only a _____limited number_____ of organisms.

7. Overproduction leads to _____competition_____ .

8. Organisms belonging to the same species can have _____different_____ traits.

9. Differences among traits are called _____variations_____ .

10. Organisms that are adapted to their environment _____reproduce_____ and pass their

    _____favorable_____ traits on to their offspring.

What evidence supports evolution?

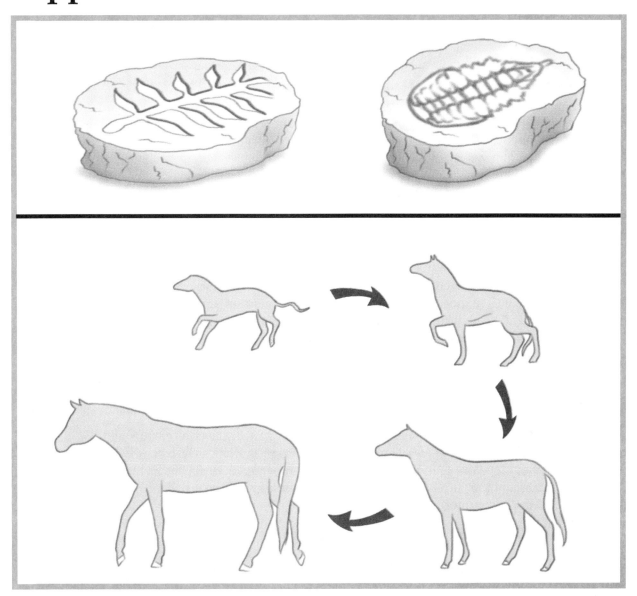

KEY TERMS

anatomy: study of the parts, or structures, of living things

vestigial structures: body parts that are reduced in size and that serve no function

LESSON 11 | What evidence supports evolution?

Imagine that you were a "spy" looking for clues to support evolution. Where would you look? There is evidence for evolution from the following different areas:

FOSSIL EVIDENCE Fossils are the remains, or traces, of organisms that lived long ago. The fossil record shows that organisms have changed over time. It shows that the earliest organisms were simple living things. They lived in water. Fossils show that these organisms evolved into more complex organisms over millions of years.

ANATOMY The study of the parts, or structures, of living things is called **anatomy** [uh-NAT-uh-mee]. By studying the parts of living things, we can find out how closely related they are. For example, the bones of a bat's wing and a human hand are similar. This suggests the animals are related.

Can you wiggle your ears? It's always good for a laugh, but nothing else, at least for modern humans. Ear movements are controlled by muscles. Human ear muscles are considered **vestigial** [ves-TIJ-ee-uhl] **structures.** Vestigial structures are "left overs." They are usually reduced in size and serve no function. Scientists think vestigial structures had a function in the ancestors of the animals that now have them. Almost all animals have vestigial structures. Humans have more than 100. The appendix is another human vestigial structure.

EMBRYOLOGY An embryo is an organism in its very early stages of development <u>before</u> it is born. Embryology is the study of embryos as they develop. Scientists compare the embryos of different living things to see if they are alike. Organisms with similar embryos probably evolved from a common ancestor.

BIOCHEMISTRY All living things are made up of chemicals called proteins. There are many kinds of proteins. Each has its own chemical "print" or structures. Scientists can identify the chemical make-up of proteins. They have discovered that the blood of certain animals have particular kinds of proteins. They compare the blood proteins of different animals. In this way, they can tell how closely the organisms are related.

- Most fossils are found in layered rocks.
- Lower layers were laid down first. They are older than the layers above them.
- Fossils found in the lower layers are older than fossils found in the upper layers.

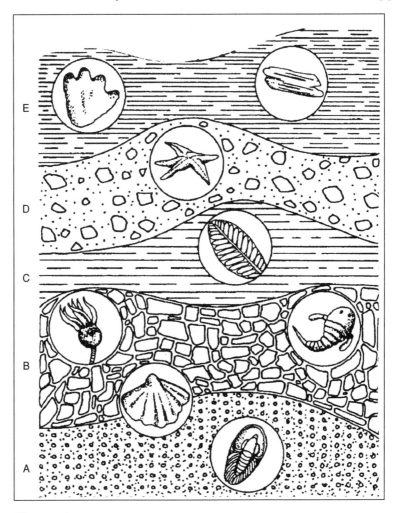

Figure A

Figure A shows five rock layers. Each contains fossils.

Study the layers and then answer the questions.

1. Which rock layer is the oldest? _____ A _____

2. Which rock layer is the youngest? _____ E _____

3. Which layer has the oldest fossils? _____ A _____

4. Which layer has the youngest fossils? _____ E _____

5. **a)** Fossils found in layer C are _____ older _____ than fossils found in layers D
   and E.   older, younger

   b) Fossils found in layer C are _____ younger _____ than fossils found in A and B.
                            older, younger

The first horse appeared about 60 million years ago. Since that time, it has been changing. Study Figure B. What changes do you see? Fill in the correct answers.

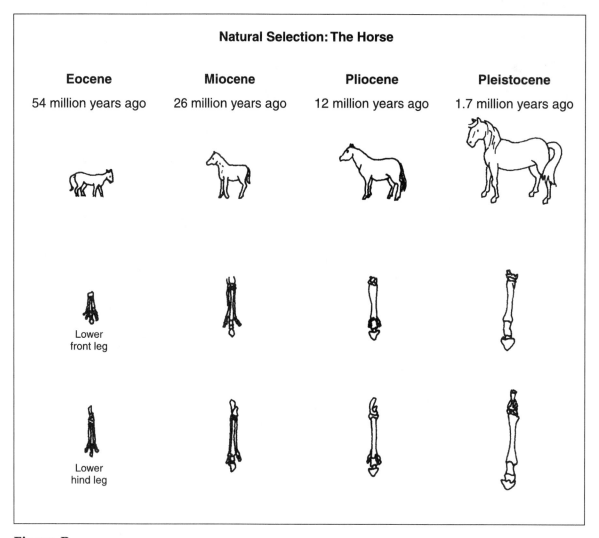

Natural Selection: The Horse

Eocene	**Miocene**	**Pliocene**	**Pleistocene**
54 million years ago	26 million years ago	12 million years ago	1.7 million years ago

Lower front leg

Lower hind leg

Figure B

1. What happened to the size of the horse? _It has gotten bigger._

2. The earliest horse had ____many____ toes.
   _{one, many}

3. How many toes does a modern horse have? ____1____ What is it called?
   (Use your own experience.) ____hoof____

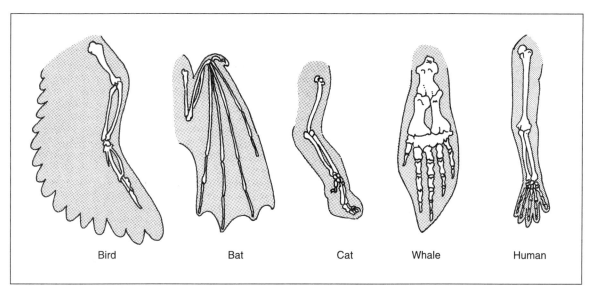

Figure C

Figure C shows the wing of a bird, the wing of a bat, the foreleg of a cat, the flipper of a whale, and the hand and arm of a human. On the outside, they look very different. Inside the bones are very similar. The bones are arranged in similar ways. They develop in much the same way.

1. Anatomy shows that these animals _____ do _____ have a close ancestor.
                              do, do not

Figure D

The wings of a bee and the wings of a bird have the same function. Both are used for flying. However, their anatomy shows that their wings are very different. They develop in totally different ways.

2. Anatomy shows that birds and bees

    are _____ distant _____ relatives.
                  distant, close

3. Birds and bees _____ did not _____
                          did, did not
    develop along the same evolutionary "branch."

The similarity of some organisms shows that they probably evolved from a common ancestor.

Figure E shows the development of a **fish**, a **turtle**, a **chicken**, a **pig**, and a **human**. Study the pictures and then answer the questions.

Fish	Turtle	Chicken	Pig	Human

Figure E

1. The adults look very _____different_____ .
   <u>similar, different</u>

2. The earliest embryos look very _____similar_____ .
   <u>similar, different</u>

___C___   **3.** Which organisms are the <u>most</u> closely related?
   a) chickens and humans   **b)** fish and pigs
   c) pigs and humans   **d)** turtles and pigs

___C___   **4.** Which organisms are the <u>least</u> closely related?
   a) fish and turtles   **b)** pigs and chickens
   c) fish and humans   **d)** pigs and humans

5. Embryos that are <u>most</u> alike are those that are the _____most_____ closely related.
   most, least

6. Embryos that are <u>least</u> alike are those that are the _____least_____ closely related.
   most, least

TRUE OR FALSE

In the space provided, write "true" if the sentence is true. Write "false" if the sentence is false.

__False__   1. Fossils found in upper rock layers are older than fossils found in lower layers.

__False__   2. Structure means how something is used.

__True__   3. Function means how something is used.

__False__   4. Different animals with parts that have similar structure and function are probably distant relatives.

__False__   5. Embryology is the study of adult organisms.

__True__   6. Closely related embryos look more alike—and for a longer time—than embryos of distant relatives.

__False__   7. The wings of bees and birds are very similar.

__True__   8. Vestigial organs have no functions.

__True__   9. Blood proteins can show evolutionary relationships.

NOW TRY THIS

Read each statement. Indicate whether each statement uses anatomy (A), biochemistry (B), or embryology (E) as evidence of evolutionary relationships among organisms. Write the correct letter in the space provided.

__A__   1. The forelimbs of a penguin and an alligator have similar bone structures.

__E__   2. The early stages of development in a fish, a rabbit, and a gorilla look alike.

__A__   3. In the wing of a bat and the arm of a human, you find bones called the radius, humerus, and ulna.

__B__   4. Some blood proteins are found in almost all organisms.

__A__   5. The finger bones in mammals have the same structure.

Complete each statement using a term or terms from the list below. Write your answers in the spaces provided. Some words may be used more than once.

fossil   layered   function
blood   vestigial   development
four   proteins   ancestor

1. In humans, the tailbone is a _____vestigial_____ structure.

2. An embryo is an organism in its early stages of _____development_____ .

3. All living things have chemicals called _____proteins_____ .

4. The wings of birds and bats show that these organisms probably have a common _____ancestor_____ .

5. Most fossils are found in _____layered_____ rocks.

6. Organisms with similar embryos probably evolved from a common _____ancestor_____ .

7. The _____blood_____ of certain animals has particular kinds of proteins.

8. Vestigial structures have no _____function_____ .

9. The earliest horse had _____four_____ toes.

REACHING OUT

The appendix is a small outgrowth at the lower part of the large intestine. In some plant–eating animals, it is much larger and is important in digestion.

Sometimes, a person's appendix becomes infected and is removed. Surgeons have removed millions of appendixes. No bad side effects have been noted after its removal.

What does this prove? _Student responses will vary. Likely responses include: The_

appendix has no function in humans.

How does adaptation help species survive?

KEY TERMS

adaptation: trait of an organism that helps it live in its environment

mimicry: adaptation of an organism that protects the organism because its appearance is similar to another organism

camouflage: ability of an organism to blend in with its surroundings

LESSON 12 | How does adaptation help species survive?

An organism must be well suited to its environment in order to survive. It must be able to tolerate the climate. It must also be able to get food, protect itself from enemies, and reproduce. An organism that is well suited to its environment is said to be adapted to its environment. In Lessons 10 and 11, you learned that organisms that are adapted to their environment have a better chance of surviving and reproducing.

Any trait of an organism that helps it live in its environment is called an **adaptation** [ad-up-TAY-shun]. Adaptations make each kind of living thing able to live in its environment. They allow one kind of organism to live where others cannot.

For example, polar bears live unprotected in subzero weather. You cannot. Neither can most organisms. Polar bears are "built" for the bitter cold. They have a thick layer of fat and dense fur to keep them warm.

Some animals live comfortably in other climates. The camel, for example, is suited to live in the hot, dry desert. Alligators are suited to live in hot, humid marshes.

Adaptation applies not only to animals. It applies to plants and other groups of living things as well. For example, a cactus can grow in the hot, dry desert. An oak tree, on the other hand, grows well in a cooler, moister environment.

Earth is about 4½ billion years old. Life has existed here for more than one billion years. Fossil records tell us that Earth has been constantly changing. So have its life forms.

Many species that lived in the past have died out. They are extinct. These organisms could not adapt to their changing environments.

Adaptation may take many forms. Here are a few examples.

The woodpecker is well adapted to dig insects out of trees.

Figure A

Birds have feathers and lightweight bones. They are well adapted for flight.

Can you think of a better way to escape an attacker? Fly away and live to fly another day.

Figure B

The streamlined shape of fish allows them to move quickly through the water.

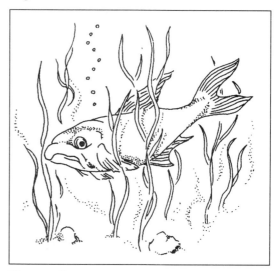

Figure C

Some organisms adapt by looking like other organisms. For example, the monarch butterfly (top) is "bad tasting."

The viceroy butterfly (bottom) is not bad tasting. Hunting birds do not know this. They are fooled by the look-alike butterfly. So they stay away from both species.

This adaptation is called **mimicry** [MIM-ik-ree].

Figure D

Some organisms blend in with their surroundings. This makes it difficult for other organisms to see them.

See how this toad blends in with its surroundings. It is almost invisible.

Figure E

The polar bear and the snowy owl blend in with the white arctic snow.

This adaptation is called **camouflage** [KAM-uh-flahj].

Figure F

The beaks and feet of birds are examples of adaptation.

Look at the diagrams of the feet and beaks of different kinds of birds. Then see if you can match the diagrams with the descriptions that follow. Place the correct letter on the lines provided.

G	**H**	**I**	**J**

K	**L**	**M**	**N**

<u> I </u>   **1.** foot adapted to swimming

<u> K </u>   **2.** foot adapted to catching and killing prey

<u> J </u>   **3.** beak adapted to spearing fish from shallow water

<u> M </u>   **4.** beak adapted to digging insects out of wood

<u> G </u>   **5.** foot adapted to clinging to branches

<u> L </u>   **6.** foot adapted to wading through shallow water

<u> H </u>   **7.** beak adapted to tearing meat

<u> N </u>   **8.** beak adapted to cracking seeds

<u> J, L </u>   **9.** The heron is classified as a wading bird. It feeds on fish and other small animals in muddy water. Which two parts of the diagram show adaptations that help the heron live in its environment?

<u> G, N </u>   **10.** The cardinal is a popular songbird of North America. It is often found perching on tree branches. Cardinals feed on wild seeds, wild fruit, and small insects. Which parts of the diagram show adaptations that help the cardinal live in its environment?

The list on the left contains words that you have used in this Lesson. Find and circle each word where it appears in the box. The spellings may go in any direction: up, down, left, right, or diagonally.

adapt

blend

cactus

camouflage

extinct

mimicry

species

suited

survive

viceroy

What are the characteristics of primates?

KEY TERMS

primates: order of mammals

opposable thumb: a thumb that can touch all of the other fingers

bipedal: upright; walk on two legs instead of four

LESSON 13 | What are the characteristics of primates?

What do tree shrews, lemurs, monkeys, apes, and humans have in common? They are all **primates**. Primates make up an order of mammals.

Early primates lived in trees. Most modern primates also live in trees. So, does it surprise you that primates have many characteristics that help them live in trees? It should not. These characteristics are adaptations to life in the trees.

All primates have flexible fingers and toes. Their fingers and toes have nails instead of claws. Some primates use their movable fingers and toes to grasp branches.

Most primates also have **opposable thumbs**. An opposable thumb can touch all of the other fingers. Touch all your other fingers with your thumb. You can see that humans have opposable thumbs too.

All primates have eyes in the front of their head. This lets them look at an object with both eyes at the same time.

Primates also have large brains compared to their body size. A large brain does not always mean greater intelligence. But, a large brain may have helped early primates adapt and survive.

Humans have all of the characteristics of other primates. They also have some characteristics that set them apart.

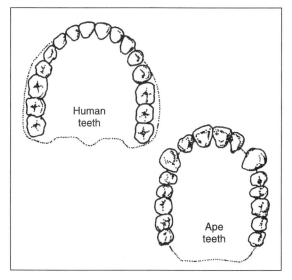

The jaws and teeth of humans are different from other primates. The human jaw is more rounded in shape. Apes have large canine teeth, also called eye teeth. Human canine teeth are not larger than other human teeth.

Figure A

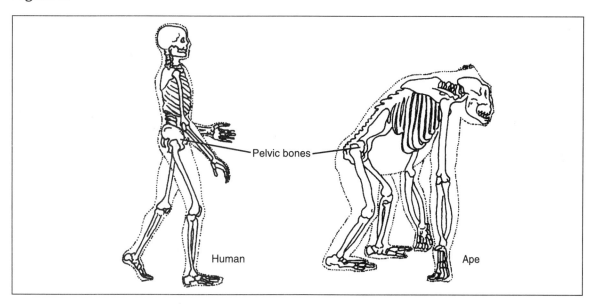

Figure B

Humans are **bipedal** [by-PEED-uhl]. Being bipedal means that humans stand upright. They walk on two legs instead of four.

Look at Figure B. You can see that an ape's pelvic bones are adapted for walking on all fours. The human pelvis allows humans to walk on two legs.

Figure C

Humans have large brains. The human brain is more highly developed than the brains of other primates.

The large front part of the brain is responsible for spoken language. Humans are the only animals that use spoken language to communicate with one another.

NOW TRY THIS

*Write **H** if the characteristic refers only to humans. Write **P** if the characteristic refers to humans and most other primates.*

___P___ **1.** Opposable thumb

___P___ **2.** Frontal vision

___H___ **3.** Bipedal

___P___ **4.** Flexible fingers

___P___ **5.** Toes with nails instead of claws

___H___ **6.** Walking upright on two legs

___H___ **7.** Rounded jaw

___H___ **8.** Spoken language

PRIMATE OR HUMAN?

*Compare the skulls in Figures D and E and the jaws in Figures F and G. Decide which is a human skull and jaw and which is an ape skull and jaw. Write either **primate** or **human** on the lines provided. Then give reasons for your answers.*

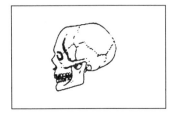

Figure D

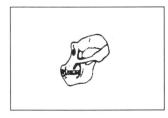

Figure E

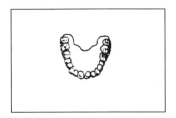

Figure F

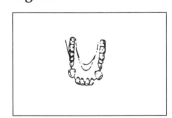

Figure G

D. ___human___

E. ___primate___

F. ___human___

G. ___primate___

Reasons: ___Humans have larger brains, so they have larger skulls. Humans have more rounded jaws than apes. Apes have larger canine teeth.___

82

How important is your thumb? More important than you may realize.

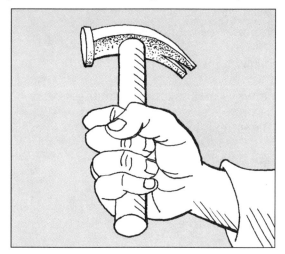

Figure H

Tape your right thumb (if you are right-handed) to your next finger. Then try to do these tasks.

1. Fasten (or unfasten) a button.

2. Pick up a book.

3. Turn a doorknob.

4. Hold an object as you would a hammer.

5. Turn a screwdriver.

How important is the thumb? <u>You</u> answer the question!

Do you think civilization would be as advanced if people did not have thumbs?

Answers will vary. Accept all logical responses.

Explain your answer. _____ Answers will vary. Accept all logical responses.

TRUE OR FALSE

In the space provided, write "true" if the sentence is true. Write "false" if the sentence is false.

_True___   **1.** Most modern species of primates live in trees.

_False___   **2.** Primates have claws instead of nails.

_False___   **3.** A large brain always means greater intelligence.

_True___   **4.** All primates have flexible fingers and toes.

_False___   **5.** All primates are bipedal.

_False___   **6.** Only humans have frontal vision.

_True___   **7.** Early primates lived in trees.

_True___   **8.** Most primates have an opposable thumb.

_False___   **9.** Apes and humans are the only animals that use sounds to communicate.

_True___   **10.** Tree shrews and lemurs are primates.

WORD SCRAMBLE

Below are several scrambled words you have used in this Lesson. Unscramble the words and write your answers in the spaces provided.

1. BINRA   _____BRAIN_____

2. ERTSE   _____TREES_____

3. PESA   _____APES_____

4. MHTBU   _____THUMB_____

5. SPERTIMA   _____PRIMATES_____

How did humans evolve?

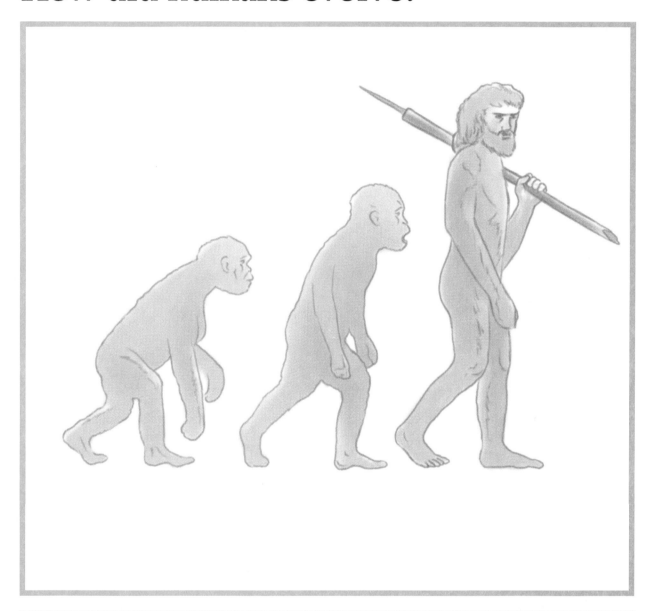

KEY TERMS

hominids: group of primates in which modern humans and their ancestors are classified

anthropologists: scientists who study human beings and trace their evolution

LESSON 14 | How did humans evolve?

Apes, chimpanzees, monkeys, tree shrews—in Lesson 13 you learned that all of these animals are primates. Humans are primates too. Modern humans and their ancestors are classified in a group of primates called **hominids** [HOM-uh-nids]. Humans are the only hominid species existing today.

The fossil record of human evolution is not complete. **Anthropologists** [an-thruh-PAHL-uh-jists] are still looking for clues to human evolution. Anthropologists are scientists who study human beings and trace their evolution.

The oldest fossils of hominids have been found in Africa. These fossils show that the earliest homonids walked upright and were about 1 meter (3 feet) tall. The fossils range from about 2½ to 3½ million years old.

Later humanlike fossils also have been found. Fossils of each species show more humanlike traits and behaviors than the species that lived before them. For example, they show an increase in body size. They have larger skulls too. Some of the later species used tools. Others lived in caves, used fire, and hunted for food.

Modern humans belong to the species *Homo sapiens* [SAY-pee-uhns]. This means "wise human." Fossils of two earlier types of modern humans have been found. They are called Neanderthals [nee-AN-dur-thals] and Cro-Magnons [kroh-MAG-nunz]. You will learn more about these two types of *Homo sapiens* on the following pages.

The first fossils of *Homo sapiens* were found in the Neander Valley of Germany. These were called Neanderthals [nee-AN-dur-thawls]. The Neanderthals lived from 130,000 to 35,000 years ago. They lived during the Ice Age in Europe.

Neanderthals were somewhat shorter than modern humans. They walked upright. They had large skulls with sloping foreheads and heavy brow bones. They had large brains.

Figure A

Their large brains helped Neanderthals adapt to the cold Ice Age. They lived in caves and used fire to keep warm. The Neanderthals also were the first people known to bury their dead.

CRO-MAGNONS

The fossils of Cro-Magnons [kroh-MAG-nuhnz] were found in a cave in France. These fossils are about 35,000 years old.

Cro-Magnons had high foreheads and no brow ridges. They looked like modern humans.

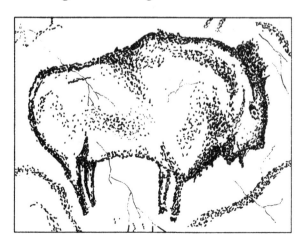

Scientists have found evidence that Cro-Magnons were skilled hunters and toolmakers. They also made sculptures and paintings on the walls of their caves.

Figure B

Decide whether the statements below refer to a Cro-Magnon or a Neanderthal. In the spaces provided, write C if the statement refers to a Cro-Magnon or N if it refers to a Neanderthal.

___N___ **1.** sloping foreheads and heavy brow bones

___C___ **2.** looked like modern humans

___N___ **3.** first *Homo sapiens* known to bury their dead

___N___ **4.** shorter than modern humans

___C___ **5.** made paintings on cave walls

Answer the following.

1. What does *"Homo sapiens"* mean? ___wise human___

2. What is anthropology? ___Study of human beings and their evolution.___

3. How do you think scientists discovered that Cro-Magnons were skilled hunters

and toolmakers? ___Answers will vary. Accept all logical responses.___

_____

4. Name two types of early humans. ___Cro-Magnons; Neanderthals___

_____

We know about human development from studying fossil bones.

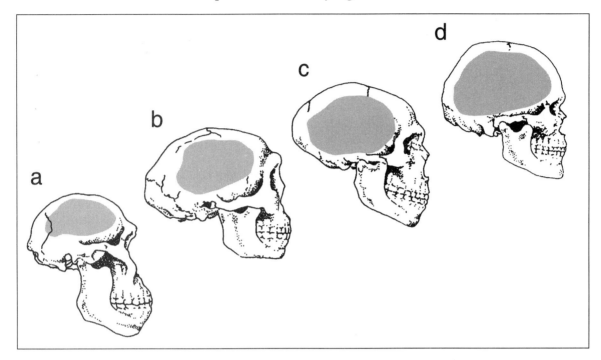

Figure C

Figure C shows four stages of human skull development—from the first man-ape to modern human.

Imagine you are an anthropologist. Compare skulls carefully. Then answer the questions below.

As humans developed:

1. Their brain size _____increased_____ .

increased, decreased

2. The skull became _____larger_____ .

larger, smaller

3. The skull also became _____more rounded_____ .

less rounded, more rounded

4. The jaw moved _____back_____ .

forward, back

5. The bony brow ridge _____became smaller_____ .

extended even more, became smaller

Study the diagrams below and then answer the questions.

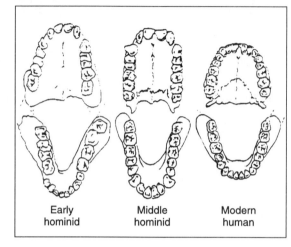

Early   Middle   Modern
hominid   hominid   human

Figure D   **Figure E**

1. As humans developed,

    a) <u>posture</u> became _____ more _____ erect.
    more, less

    b) the <u>head</u> dropped _____ less _____ .
    more, less

2. The <u>body</u> had _____ less _____ hair.
    more, less

3. The "<u>eye</u>" <u>teeth</u> (canines) of modern humans are _____ smaller _____ than those of
    smaller, larger
    earlier hominids.

4. The <u>mouth</u> (not lips) became more _____ rounded _____ .
    "U" shaped, rounded

5. <u>Reasoning power</u> _____ improved _____ .
    improved, remained the same, became worse

6. <u>Ability to communicate</u> _____ improved _____ .
    improved, remained the same, became worse

7. <u>Coordination</u> _____ improved _____ .
    improved, remained the same, became worse

8. There is <u>one</u> main reason for your answers to questions 5, 6, and 7.

    What is that reason? _____ larger, more developed brain _____

Four skulls and four heads are shown below. They are not matched. Match each skull to its correct head. Write your answers in the chart below.

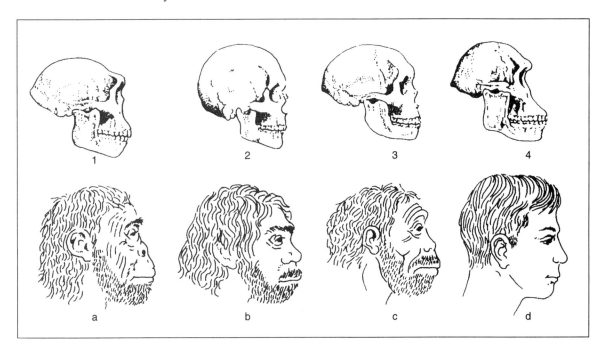

Figure F

SKULL	MATCHING HEAD
1	C
2	D
3	B
4	A

1. Now arrange the pairs according to <u>advancing</u> evolution.

   (Earliest pair first) ___4 + A, 1 + C, 3 + B, 2 + D___

2. Which was the least developed skull? ___4___

3. How do you know? ___Accept all logical student explanations.___

4. Which was the most developed skull? ___2___

5. How do you know? ___Accept all logical student explanations.___

FILL IN THE BLANK

Complete each statement using a term or terms from the list below. Write your answers in the spaces provided. Some words may be used more than once.

most advanced	primates	Germany
largest	complete	hominids
fossil bones	anthropologists	humans

1. Apes, chimps, monkeys, and humans are classified as ___primates___ .

2. Primates are the ___most advanced___ of all animals.

3. The most intelligent primates are ___humans___ .

4. Relative to their size, humans have the ___largest___ brain.

5. Scientists who study human development are called ___anthropologists___ .

6. Modern humans and their ancestors are classified as ___hominids___ .

7. The only surviving hominids are ___humans___ .

8. The fossil record of human evolution is not ___complete___ .

9. Most of what we know about human evolution has come from studying

   ___fossil bones___ .

10. The first fossils of modern humans were found in ___Germany___ .

WORD SCRAMBLE

Below are several scrambled words you have used in this Lesson. Unscramble the words and write your answers in the spaces provided.

1. SNAHUM ___HUMANS___

2. CEVAS ___CAVES___

3. PSIMRTAE ___PRIMATES___

4. SHOINIMD ___HOMINIDS___

5. LKUSL ___SKULL___

TRUE OR FALSE

In the space provided, write "true" if the sentence is true. Write "false" if the sentence is false.

___True___   **1.** The oldest fossils of hominids have been found in Africa.

___False___   **2.** Hominids include apes, monkeys, and humans.

___False___   **3.** The earliest hominids were as tall as modern humans.

___True___   **4.** *Homo sapiens* means "wise human."

___True___   **5.** All modern humans belong to the species *Homo sapiens.*

___False___   **6.** Neanderthal fossils were first found in a cave in France.

___True___   **7.** Cro-Magnons were skilled hunters and toolmakers.

___True___   **8.** Neanderthals lived during the Ice Age.

___False___   **9.** Cro-Magnons were the first people known to bury their dead.

___True___   **10.** Fossils of each hominid species show more humanlike traits than the species that lived before them.

REACHING OUT

The words *Homo sapien* mean "wise human." Why do you think scientists gave this

name to this group of hominids? ____Answers will vary. Accept all logical responses.____

_____

_____

_____

SCIENCE *EXTRA*

Cloning

For hundreds of years, people have raised useful animals. They have bred horses that can pull heavy loads. They have also bred dogs that can hunt and cattle that taste good. Now, scientists have found a new way to raise animals. It is called cloning. Through cloning, people can raise identical copies of an animal.

In normal animal reproduction, animals mate to produce offspring. A sperm from a male combines with an egg from a female. The offspring gets half its DNA from its mother and half from its father. The new animal is similar to each parent in some ways. However, it isn't identical to either one.

In cloning, scientists use DNA from only one animal. The animal may be male or female. They place the DNA into an egg whose own DNA has been removed. If the scientist is cloning a mammal, the egg is then placed into the uterus of an adult female of the same species. The egg grows into a new animal just as any baby animal would. However, this baby is identical to the animal that supplied the DNA.

Scientists think cloning may solve some health problems. For example, an animal may have a genetic disease normally found in humans. Scientists could clone that animal. Then, a drug to treat the disease could be tested on the clones before it is tested on humans.

Scientists also use clones to make drugs. Some drugs are made from chemicals in animals' bodies. If a particular animal produces a useful chemical, that animal could be cloned. Each clone would be a living source for useful drugs.

Lots of questions about cloning haven't been answered yet. Will cloned animals age in the same way as other animals? Can humans be cloned? If they can, who are the parents? Scientists and lawmakers will have to work together to make sure that cloning is used safely and sensibly.

What are viruses?

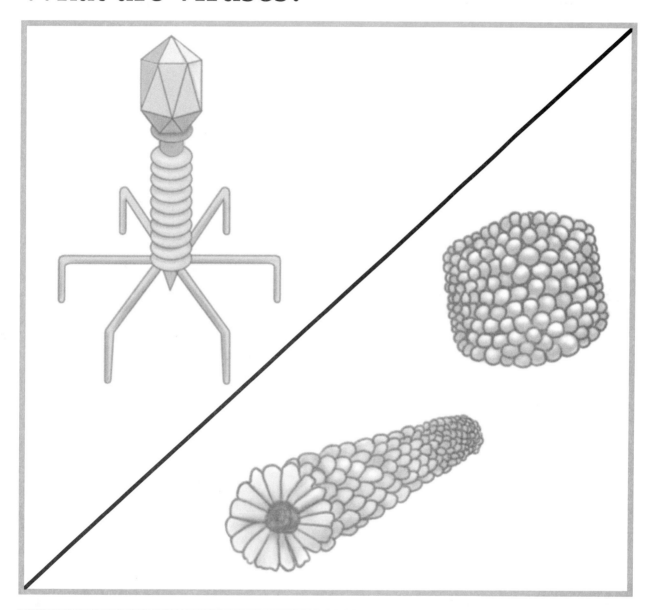

KEY TERMS

virus: piece of nucleic acid covered with an outercoat of protein

nucleic acids: organic compounds that make proteins, control the cell, and determine heredity

capsid: coat of protein that covers a virus

LESSON 15 | What are viruses?

Alive or not alive? That is the question! Scientists do not agree about whether viruses are actually living things.

Viruses are unusual. A virus has no cell parts. A **virus** is just made up of a substance called **nucleic** [new-KLEE-ik] **acid** covered by an "overcoat" of protein. The outer coat is called a **capsid** [KAP-sid]. The capsid makes up most of the virus.

Capsids give viruses their shapes. Some viruses are round. Others look like long rods. Some have very unusual shapes. You can see the shapes of some viruses in Figure A on the facing page.

How else does a virus differ from living cells? A virus does not ingest or digest food. It does not carry out respiration. In fact, a virus does not carry out any of the life processes except reproduction—and then, only when it is inside a living cell. When a virus is outside a living cell, it is just a "chemical." What happens when a virus infects a cell? It may cause disease. Viruses cause many diseases in plants and animals. When you have the "flu," you are infected with a virus. Viruses also infect bacteria.

Because viruses do not have all the characteristics of living things, they are not classified in the five kingdoms. Instead, most scientists classify viruses according to the living things they infect. The three main groups are: plant viruses, animal viruses, and bacterial viruses.

Viruses are ultra-small. Bacteria are tiny. Yet, compared to viruses, bacteria are giants! Think of this . . . a single microscopic cell may contain millions of virus particles.

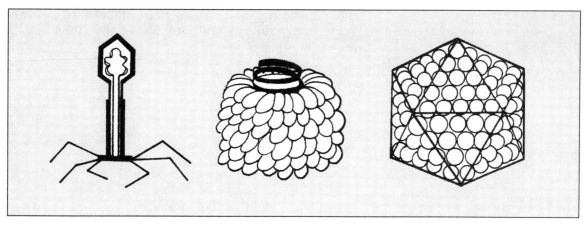

Figure A

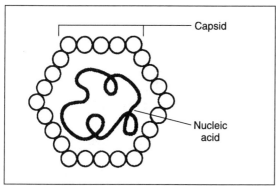

Figure B

1. The two parts of a virus are

   _____nucleic acid_____ and a _____capsid_____ .

2. Which part makes up most of a virus?

   _____capsid_____

3. Which part gives a virus its shape?

   _____capsid_____

4. How do viruses differ from living cells? _____A virus has no cell parts and it_____
   _____does not carry out the life processes._____

5. What is the only life function that a virus carries out? _____reproduction_____

6. When is the only time this life process can take place? _____when a virus is inside_____
   _____another living cell_____

7. How do most scientists classify viruses? _____according to the living things_____
   _____they affect_____

8. What are the three main groups of viruses? _____plant viruses, animal viruses,_____
   _____and bacterial viruses_____

When a virus enters a cell, it takes over the cell and causes it to make new viruses. Scientists first learned about how viruses reproduce by studying bacterial viruses.

As you read about how viruses reproduce, look at Figure C.

1. A virus attaches to a cell.

2. The virus sends its nucleic acid into the cell. The capsid stays outside.

3. The virus takes over the cell. It directs the cell to make new viruses.

4. The new viruses burst out of the cell. This kills the cell. The new viruses attack other cells.

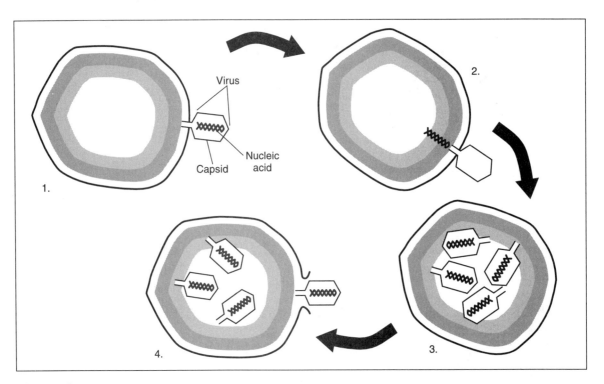

Figure C

FILL IN THE BLANK

Complete each statement using a term or terms from the list below. Write your answers in the spaces provided. Some words may be used more than once.

diseases	reproduction	cell
shapes	infect	capsid
nucleic acid	animal	smaller

1. Scientists classify viruses according to the living things they _____infect_____ .

2. A virus is just a piece of _____nucleic acid_____ covered by protein.

3. Some viruses have very unusual _____shapes_____ .

4. The outer coat of a virus is called a _____capsid_____ .

5. Capsids give viruses their _____shapes_____ .

6. The three main groups of viruses are bacterial, plant, and _____animal_____ viruses.

7. Viruses are much _____smaller_____ than bacteria.

8. A virus does not carry out any of the life processes except _____reproduction_____ .

   However, it can reproduce only inside a living _____cell_____ .

9. A virus has no _____cell_____ parts.

10. Viruses cause many plant and animal _____diseases_____ .

TRUE OR FALSE

In the space provided, write "true" if the sentence is true. Write "false" if the sentence is false.

__True__ **1.** The flu is caused by a virus.

__True__ **2.** The capsid makes up most of a virus.

__False__ **3.** All viruses are round.

__False__ **4.** When viruses reproduce, the capsid enters a cell.

__True__ **5.** The outer coat of a virus is made up of protein.

__False__ **6.** The nucleic acid gives a virus its shape.

__True__ **7.** Viruses do not ingest or digest food.

__False__ **8.** Scientists first learned about how viruses reproduce by studying animal viruses.

Match each statement to the stages showing how viruses reproduce. Write the letter of each statement below the correct stage.

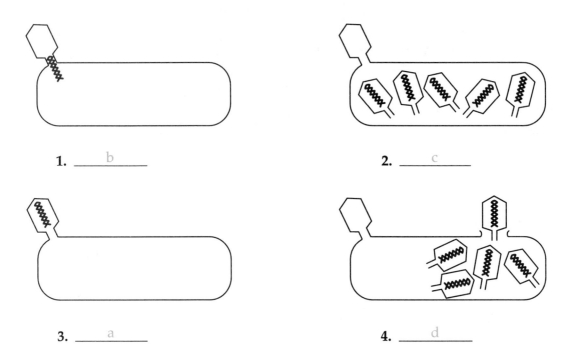

1. ___b___

2. ___c___

3. ___a___

4. ___d___

Figure D

a. Bacterial virus attaches itself to the host cell.
b. Virus injects its nucleic acid into the cell.
c. Nucleic acid of virus directs the cell to make new viral nucleic acid and capsids.
d. New viruses burst out of the host cell.

REACHING OUT

Do you think viral infections are hard to treat? Explain why or why not.

Answers will vary. Accept all logical responses.

Lesson 16

What are infectious diseases?

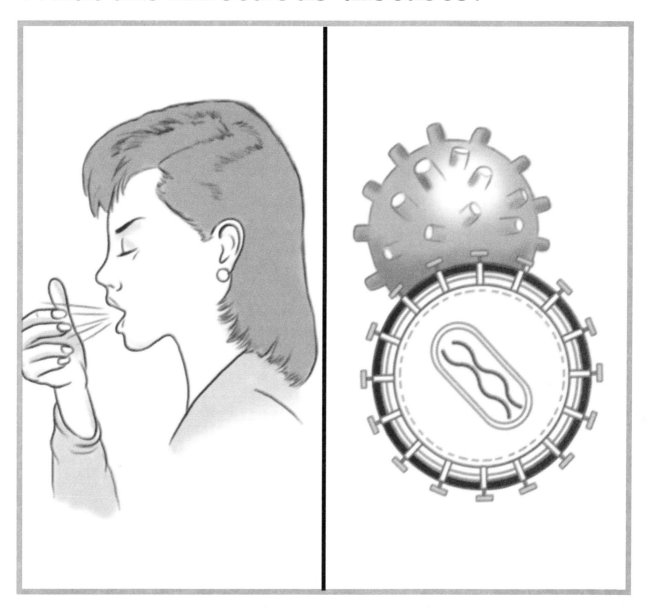

KEY TERMS

infectious disease: disease caused when a virus (or germs) enter the body

toxins: poisons

contagious disease: infectious disease that can be transmitted from person to person

immune system: body system made up of cells and tissues that help a person fight disease

AIDS: viral disease that attacks a person's immune system

LESSON 16 | What are infectious diseases?

Disease is a great enemy of living things. A disease interferes with the normal functions of an organism.

Diseases injure or kill millions of people each year. Diseases also infect plants and other animals. They destroy crops. They kill animals such as cattle and chickens.

Some diseases are caused by <u>germs</u> that enter the body. The germs may be bacteria, other microscopic organisms, or viruses. These diseases are called **infectious** [in-FEK-shus] **diseases**. The "flu" is an example of an infectious disease. As you learned in Lesson 16, it is caused by a virus.

Disease symptoms develop when microbes destroy cells. Most disease-causing bacteria produce poisons called **toxins**. These toxins kill the cells. Some microbes live in cells. They multiply so fast that the cells die.

Some infectious diseases can be spread from one person to another. These kinds of infectious diseases are **contagious** [kuhn-TAY-jus] **diseases**. Have you ever heard someone with a cold say, "Don't come close. I might be contagious"? The common cold is a contagious disease.

Now let us examine some of the ways diseases are spread.

- Some diseases are spread through the air. When an infected person coughs or sneezes, germs are sprayed into the air.

- Many diseases are spread by taking in food or water that contains germs.

- Some are spread by contact with objects that have germs.

- The bite of an insect spreads certain diseases.

- Some diseases are spread by direct contact with an infected person.

SOME COMMON INFECTIOUS DISEASES

The chart below lists several infectious diseases of humans. Study the chart. Then answer the questions on the next page.

DISEASE	CAUSED BY	SYMPTOMS
Chicken pox	Virus	fever, headache, skin rashes that form crusts
Measles	Virus	fever, rash, sensitivity to light, cough, body aches
Botulism [BAHT-yoo-lizm]	Bacteria	double vision, abdominal pain, heart and lung paralysis
Malaria	Protozoa	fever, chills
Strep Throat	Bacteria	fever, sore throat
Mumps	Virus	fever, chills, headache, swollen neck and throat glands
Influenza [in-floo-EN-zuh] (flu)	Virus	fever, chills, body aches, [and possibly a sore throat and cough]
Athlete's foot	Fungus	itchy skin
Polio	Virus	fever, sore throat, stiff back, muscle pain, paralysis
Tetanus	Bacteria	tightening of muscles

1. Which of these common diseases have you had? _____Answers will vary._____

2. What is the most common symptom of these infectious diseases? _____
   ___fever_____

3. a) What are the symptoms of malaria? ___fever and chills_____

   _____

   b) What causes malaria? ___a protozoan_____

4. Which diseases listed in the chart are caused by bacteria? ___botulism, strep___
   ___throat, tetanus_____

5. Which of the diseases are caused by a virus? ___chicken pox, measles, mumps,___
   ___influenza, polio_____

MATCHING

Match the disease listed in Column A to its symptoms in Column B.

	Column A	Column B
___c___	1. polio	a) fever, chills, swollen neck and throat glands
___b___	2. chicken pox	b) fever, headache, skin rashes that form crusts
___e___	3. athlete's foot	c) fever, sore throat, stiff back, muscle pain, paralysis
___d___	4. tetanus	d) tightening of muscles
___a___	5. mumps	e) itchy skin

Farm animals like cattle, hogs, and sheep are called livestock.

Animal diseases kill more than two billion dollars worth of livestock each year in the United States alone.

One of the worst livestock diseases is hoof-and-mouth disease. It is caused by a virus and it spreads very rapidly. Many of the infected animals die.

A farmer may have to kill all the animals that are near an infected animal to prevent an epidemic. An epidemic is the spread of a disease to many organisms in an area at the same time.

Figure A

PLANT DISEASES

At one time, the most important food crop in Ireland was the potato.

During the 1840s, a fungus disease destroyed the potato crop in Ireland. Between 1845 and 1847, nearly 750,000 people died of starvation. Hundreds of thousands of others fled the country in search of food and a new life. Many of these people came to the United States.

Figure B

FILL IN THE BLANK

Complete each statement using a term or terms from the list below. Write your answers in the spaces provided.

contagious	toxins	coughs
spread	virus	infectious
insect	hoof-and-mouth	epidemic
sneezes	fungus	

1. Contagious diseases can be _____ spread _____ from one person to another.

2. One of the worst livestock diseases is _____ hoof-and-mouth _____ disease.

3. Diseases caused by germs are called _____ infectious _____ diseases.

4. The bite of an _____ insect _____ spreads some diseases.

5. The spread of a disease to many organisms in an area is called an _____ epidemic _____ .

6. Most disease-causing bacteria produce _____ toxins _____ .

7. The potato famine of Ireland in the 1840s was caused by a _____ fungus _____ .

8. The "flu" is caused by a _____ virus _____ .

9. When an infected person _____ coughs _____ or _____ sneezes _____ germs are sprayed into the air.

10. The common cold is a _____ contagious _____ disease.

TRUE OR FALSE

In the space provided, write "true" if the sentence is true. Write "false" if the sentence is false.

_____ False _____ 1. The common cold is not a contagious disease.

_____ False _____ 2. Athlete's foot is caused by bacteria.

_____ True _____ 3. Diseases injure or kill millions of people each year.

_____ False _____ 4. Infectious diseases affect only animals.

_____ True _____ 5. Germs may be bacteria, viruses, protozoa, or fungi.

_____ False _____ 6. All infectious diseases are caused by viruses.

_____ False _____ 7. Hoof-and-mouth disease is one of the worst human diseases.

_____ True _____ 8. Malaria is an infectious disease.

_____ True _____ 9. Many diseases are spread by taking in food or water that contains germs.

_____ True _____ 10. Covering your mouth when you cough helps prevent the spread of germs.

106

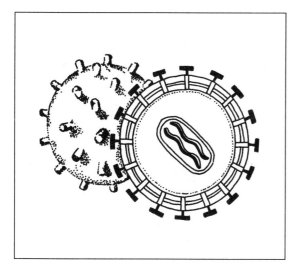

You know that viruses cause many diseases. One particular virus called the <u>HIV</u> virus attacks a person's **immune system**. Your immune system is made up of cells and tissues that fight disease.

Figure C *HIV virus*

The disease caused by the HIV virus is called <u>Acquired Immune Deficiency Syndrome</u>, or **AIDS**.

Because the HIV virus attacks the immune system, a person with AIDS loses the ability to fight disease. The person gets diseases that most healthy people can fight off.

HOW IS AIDS SPREAD?

People with AIDS have the HIV virus in their blood and body fluids. In order to be infected with the HIV virus, you must exchange bodily fluids with an infected person.

The HIV virus can enter the bloodstream by sexual contact with someone who has AIDS. It can also be spread by intravenous drug users who use contaminated needles. A third way the HIV virus enters the bloodstream is through a blood transfusion of infected blood. However, most blood in the United States is tested for the HIV virus.

AIDS TREATMENT

There is no known cure for AIDS at this time. It is a fatal, or deadly, disease.

Scientists are working on ways to treat AIDS patients and make their immune systems stronger. They are also working on an AIDS vaccine.

WAYS OF AIDS INFECTION

Place a check mark beside each statement that describes a way that a person can become infected with the AIDS virus.

_____ **1.** from a dog bite

_____✔_____ **2.** exchange of body fluids with an infected person

_____✔_____ **3.** sexual contact with an infected person

_____ **4.** from a mosquito bite

_____ **5.** through casual contact with an infected person

_____✔_____ **6.** by a contaminated needle

Complete the following.

1. What is AIDS? _____disease of the immune system caused by the HIV virus_____

2. What do the letters in the word "AIDS" stand for? _____Acquired Immune Deficiency Syndrome_____

3. What is the name of the virus that causes AIDS? _____HIV Virus_____

4. What is the immune system? _____cells and tissues that fight disease_____

5. Why are diseases that a healthy person can fight off sometimes fatal to a person with AIDS? _____A person who has AIDS has a weakened immune system._____

What are noninfectious diseases?

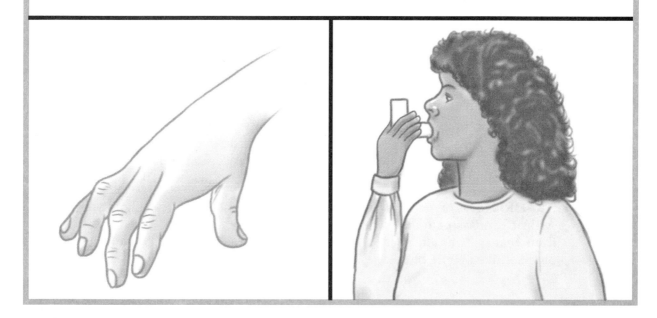

SURGEON GENERAL'S WARNING: Quitting Smoking Now Greatly Reduces Serious Risks to Your Health.

KEY TERM

noninfectious diseases: diseases that are not caused by germs and not spread from person to person

LESSON 17 | What are noninfectious diseases?

In Lesson 16, you learned that infectious diseases are caused by germs. But what about other diseases such as heart disease? These are diseases too. However, they are not caused by germs. They are called **noninfectious diseases**.

A noninfectious disease is not spread from person to person. Some noninfectious diseases last for a long time or keep coming back. These kinds of diseases are called chronic illnesses. Cancer is one example of a chronic illness.

There are many groups of noninfectious diseases. Some of the major noninfectious diseases are identified below.

HEART DISEASE is the leading cause of death in the United States. Heart disease develops when the normal flow of blood through the heart or body is stopped in some way. Some common forms of heart disease are heart attacks, strokes, and high blood pressure.

CANCER occurs when certain cells in the body grow without control. The cancer cells destroy normal tissue. If left untreated, most kinds of cancer cause death. However, many forms of cancer can be treated if they are caught early enough.

ARTHRITIS is a general term for conditions that affect the joints. Arthritis causes pain and swelling in many joints of the body. You may think that arthritis affects mostly older people. Some forms do. However, others strike people of all ages.

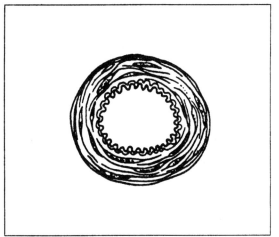

Figure A *Healthy artery*

In one kind of heart disease, fatty substances build up. They build up on the walls of the arteries. One of these fatty substances is cholesterol [kuh-LES-tuh-rohl]. Cholesterol is a fatty substance found in animal products.

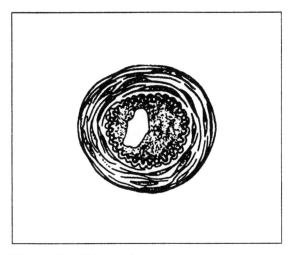

Figure B *Diseased artery*

As fat builds up in an artery, the artery becomes more narrow. This makes the heart have to work harder. The heart must work harder to pump blood through narrow arteries.

HEART ATTACKS AND STROKES

Sometimes the arteries leading to the heart are blocked. This prevents food and oxygen from reaching the heart. Then the heart cannot do its work. This is called a <u>heart attack</u>.

If the arteries to the brain are blocked, food and oxygen cannot reach the brain, this is called a <u>stroke</u>.

Do you think heart disease can be prevented? Scientists have found that there are several things that increase a person's chance of getting heart disease. Some of these things cannot be changed. However, others can be controlled.

Look at the chart below. It shows the major factors that lead to heart disease. Then answer the questions below the chart.

FACTORS CONTRIBUTING TO HEART DISEASE

• age (risk increases with age)	• obesity
• family history (runs in families)	• physical inactivity
• gender (men are at greater risk than women)	• high cholesterol levels
• smoking	• poor eating habits
• high blood pressure	

1. Which factors do you think cannot be controlled? _____ age, family history, gender _____

2. Which factors may be controlled? _____ smoking, high blood pressure, obesity, _____
   physical activity, high cholesterol levels, poor eating habits

TRUE OR FALSE

In the space provided, write "true" if the sentence is true. Write "false" if the sentence is false.

True	**1.**	The heart must work harder to pump blood through narrow arteries.
False	**2.**	Cholesterol is a fatty substance found in plant products.
True	**3.**	There are several risk factors for heart disease.
False	**4.**	When arteries to the heart are blocked, a stroke occurs.
False	**5.**	Heart attacks and strokes are the only forms of heart disease.
True	**6.**	Smoking increases your risk for heart disease.
True	**7.**	Arteries to the heart bring it food and oxygen.
True	**8.**	The risk of heart disease increases with age.
False	**9.**	You can control all the risk factors for heart disease.
False	**10.**	Women have a greater risk of heart disease than men.

MORE ABOUT CANCER

Cells in the body usually divide and grow in an orderly manner. Sometimes growth goes wild. Cells grow out of control. The cells form a mass, or lump, called a <u>tumor</u> [TOO-mur].

There are two kinds of tumors:

<u>BENIGN</u> [bi-NYN] <u>TUMORS</u> grow only in one place in the body. A benign tumor does not spread to other places. It is usually not a serious problem.

<u>MALIGNANT</u> [muh-LIG-nunt] <u>TUMORS</u> spread to other places in the body. As a malignant tumor spreads, it harms the body. If its growth is not stopped, a person with a malignant tumor may die. Cancer is the disease a person with a malignant tumor has.

Scientists are not sure what causes all types of cancer. But, they do know that certain things cause cancer or increase a person's chance of getting cancer. Some of these things are smoking, X rays, and too much sunlight.

EARLY SIGNS OF CANCER

The earlier cancer is detected, the more likely a person is to survive. The chart below list the seven warning signs of cancer. A person with any of these signs should see a doctor immediately.

1. A sore that does not heal
2. Unusual bleeding
3. A lump in the breast or other area beneath the skin
4. Constant indigestion or trouble swallowing
5. A nagging cough
6. A change in size, shape, or color of a wart or mole
7. A change in bowel or bladder habits

MATCHING

Match each term in Column A with its description in Column B. Write the correct letter in the space provided.

	Column A	Column B
c	**1.** cancer	**a)** harmless mass of cells
a	**2.** benign tumor	**b)** any mass or lump of cells
b	**3.** tumor	**c)** rapid, uncontrolled growth of cells
d	**4.** malignant tumor	**d)** harmful mass of cells that can spread throughout the body
e	**5.** sunlight	**e)** possible cause of skin cancer

113

Place a check mark beside each statement that describes one of the seven warning signs of cancer.

_____ ✔ _____ **1.** unusual bleeding

_____ _____ **2.** thirst

_____ ✔ _____ **3.** nagging cough

_____ ✔ _____ **4.** sore that does not heal

_____ _____ **5.** constant hunger

_____ ✔ _____ **6.** change in size of a mole

_____ ✔ _____ **7.** changes in color of a mole

_____ ✔ _____ **8.** lump beneath the skin

_____ _____ **9.** sleeplessness

_____ ✔ _____ **10.** constant indigestion

Study the illustrations. Place a check mark below each illustration that shows a possible cause of cancer.

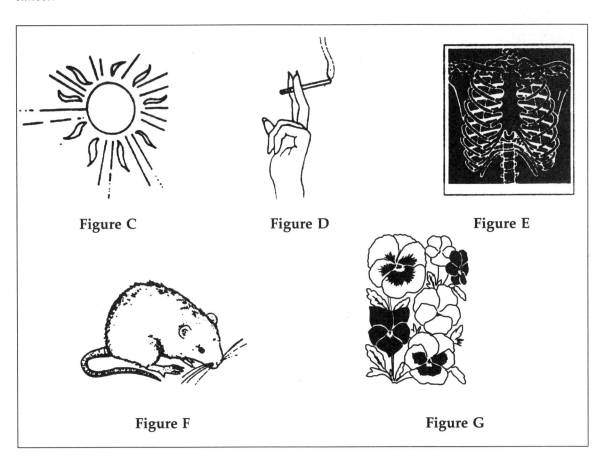

Figure C

Figure D

Figure E

Figure F

Figure G

Students should place check marks below Figures C, D, and E.

Heart disease, cancer, and arthritis are only three of the many noninfectious diseases. Other groups of noninfectious diseases are described below.

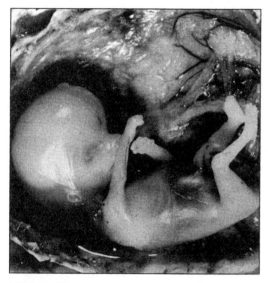

Figure H

Some diseases and disorders are present at birth. They include heart, lung, and eye problems, blood disorders, hemophilia, bone deformities, and mental retardation.

Some diseases and disorders are passed on by genes. They are inherited. However, some others are caused by an unhealthy environment before birth. For example, babies can be born with many problems if their parents took drugs.

NUTRITIONAL DISEASES

Other types of diseases are caused by an improper diet. These are called nutritional diseases. The table below list some nutritional diseases.

NUTRITIONAL DISEASES		
DISEASE	**SYMPTOMS**	**CAUSED BY A DEFICIENCY OF**
Anemia	lack of energy	Vitamins B_6, B_{12}, or iron
Scurvy	sore gums	Vitamin C
Rickets	soft bones and teeth	Vitamin D; Calcium
night blindness	difficulty seeing in dark	Vitamin A
goiter	swollen thyroid gland	Iodine

Complete each statement using a term or terms from the list below. Write your answers in the spaces provided.

fatty	joints	tumor
brain	genes	heart
cells	birth	vitamin C
narrow		

1. Arthritis is a general term for conditions that affect the ____joints____ .

2. A stroke occurs when arteries to the ____brain____ are blocked.

3. A benign ____tumor____ is usually not a serious problem.

4. The leading cause of death in the United States is ____heart____ disease.

5. Some congenital diseases are passed on by ____genes____ .

6. Congenital diseases are present at ____birth____ .

7. Cholesterol is a ____fatty____ substance.

8. The heart must work harder to pump blood through ____narrow____ arteries.

9. Scurvy is caused by lack of ____vitamin C____ in the diet.

10. Cancer occurs when ____cells____ grow and divide without control.

REACHING OUT

Many foods advertise that they have no cholesterol or are low in cholesterol. Why do you think companies use this as a

selling point? __Student answers will__

__vary. Likely responses will include:__

__People who are concerned about their__

__intake of cholesterol may find the__

__product more appealing.__

Figure I

How does the body protect itself from disease?

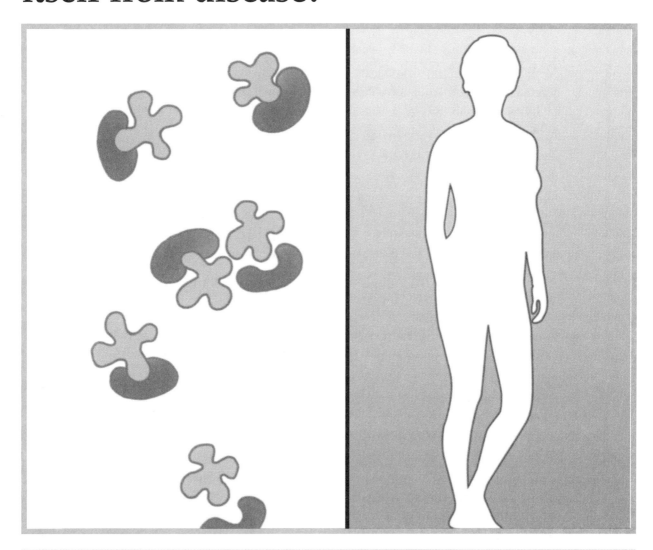

KEY TERMS

mucus: sticky substance made by the body that traps germs

cilia: tiny hairlike structures

white blood cells: cells that protect the body against disease

antibodies: proteins made by the body that destroy germs

immune system: body system made up of cells and tissues that help a person fight disease

immunity: resistance to a certain disease

LESSON 18 | How does the body protect itself from disease?

Your body is under constant attack from viruses, bacteria, and other germs. However, your body can usually protect itself from disease. It has defenses against germs.

Your skin is your body's first line of defense. The skin is a waterproof, germ-proof barrier that covers your body. The skin acts like a wall to keep out germs.

Your mouth and nose are two places where germs can enter the body. The inside of the nose is lined with small hairs and sticky liquid called **mucus**. The hairs filter out dust and pollen from the air. The mucus traps germs (usually bacteria) as well as dust and pollen.

Your windpipe, like the nose, is lined with mucus. Both your windpipe and your nose are also lined with very tiny (microscopic) hairs called **cilia** [SIL-ee-uh]. Cilia are always beating in an outward direction. The mucus traps many harmful substances. The cilia sweep them outward. Coughing and sneezing also help force germs out of the body.

Now suppose germs do get by the body's first defenses. What happens? **White blood cells** go to work. It is the job of white blood cells to destroy germs that are harmful to the body. Special white blood cells destroy germs by surrounding them and taking them inside the white blood cells where they are digested.

Your body also has one more line of defense. The body is able to make chemical substances that destroy germs. These substances are called **antibodies** [AN-ti-bahd-eez]. Antibodies clump together with germs and destroy them.

In Lesson 17, you learned that the immune system is made up of tissues and cells that fight disease. The **immune system** is in charge of recognizing germs and making antibodies.

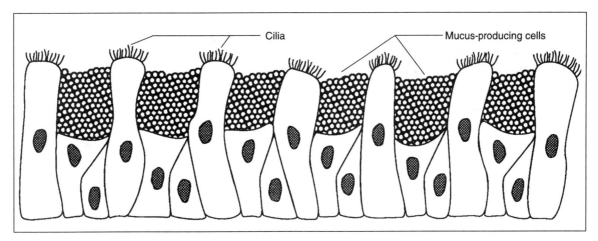

Figure A *The cilia and mucus in the nose are part of the body's first line of defense.*

1. The nose is lined with small _____hairs_____ and sticky liquid called _____mucus_____ .

2. The hairs and mucus filter and trap _____germs_____ , _____dust_____ , and _____pollen_____ .

3. Trapped dust and pollen "tickle" our noses. This makes us _____sneeze_____ .

4. How do sneezing and coughing help fight disease? _____They force germs out of the body._____

5. Why should you always "cover" a sneeze or cough? _____to prevent the spread of germs_____

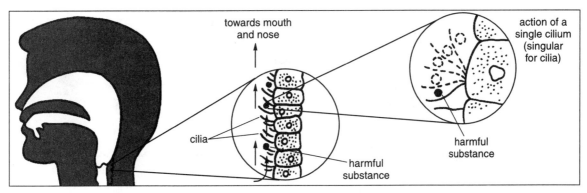

Figure B

6. Harmful substances that reach the windpipe are trapped by _____mucus_____ .

7. They are swept outward by microscopic hairs called _____cilia_____ .

8. Cilia in the windpipe are always moving towards _____the mouth and nose_____ .
    the lungs, the mouth and nose

Study the pictures below. Then answer the questions.

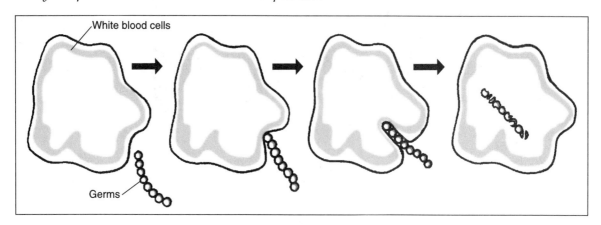

Figure C

1. What kind of blood cells fight germs in the body? _white blood cells_

2. What is happening in Figure C? _a white blood cell is surrounding and digesting_ _a germ_

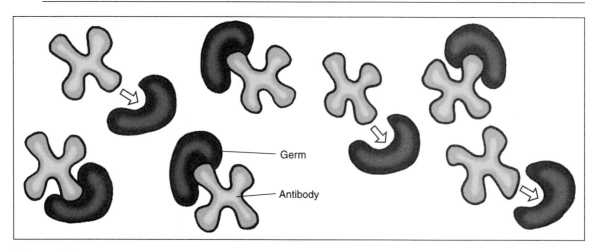

Figure D

3. What chemical substances does the body make to fight germs? _antibodies_

4. What is happening in Figure D? _antibodies are combining with germs to_ _destroy them_

5. What body system is in charge of making antibodies? _immune system_

Complete each statement using a term or terms from the list below. Write your answers in the spaces provided. Some words may be used more than once.

sticky	immune system	defenses
outward	antibodies	germs
filter	skin	surround

1. Your body has _____defenses_____ against disease.

2. The immune system produces _____antibodies_____ .

3. Cilia are always beating in an _____outward_____ direction.

4. The _____skin_____ acts like a wall to keep out germs.

5. White blood cells destroy _____germs_____ that enter the body.

6. Cells and tissues that fight disease make up the _____immune system_____ .

7. Hairs in the nose _____filter_____ air.

8. Some white blood cells _____surround_____ germs to destroy them.

9. Chemical substances called _____antibodies_____ clump together with germs and destroy them.

10. Mucus is a _____sticky_____ liquid.

COMPLETE THE TABLE

Complete the table by describing how each part of the body's defense systems helps to protect the body from harmful germs.

Body Defenses Against Disease		
	Defense	**How It Works**
1.	Skin	prevents harmful germs from entering the body
2.	Nose	hairs in the nose filter germs from air; mucus traps germs; cilia sweep out germs
3.	Cilia and Mucus	trap and sweep out germs
4.	White Blood Cells	destroy germs by digesting, or eating them
5.	Antibodies	clump together with germs and destroy them

Antibodies destroy germs. After the germs are destroyed, many of the antibodies remain. If the same kind of germs enter the body again, the antibodies are "ready and waiting." They destroy the germs before they can do any harm. The body has become resistant.

Resistance to a certain disease is called **immunity** [im-MYOON-i-tee].

There are two kinds of immunity—natural immunity and acquired [uh-KWY-urd] immunity.

- Natural immunity is immunity that you are born with. It is your body's natural defense against disease.

- Acquired immunity is immunity that you get or develop during your life.

There are several ways you can acquire or get immunity.

- You can be given a shot of antibodies against a certain disease.

- Developing babies get antibodies from their mothers.

- Once you have had some diseases, your body keeps making antibodies against that disease. Did you ever have chicken pox? If you have, you now have immunity against the chicken pox.

- You can get a vaccine.

Vaccinations are helpful in preventing specific diseases such as polio and the measles. A vaccine is made up of specific dead or weakened disease-causing microbes. However, a vaccine does not cause you to get the disease.

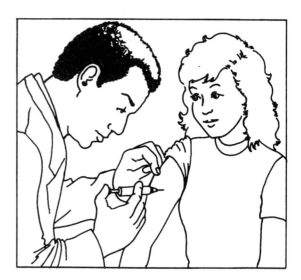

Figure E

How does a vaccine work? The vaccine enters the body by injection, through a scratch or by swallowing. It signals the body to make antibodies.

A vaccinated individual becomes resistant to a specific disease. But a "booster" shot may be required, after a period of time, to keep the immunity.

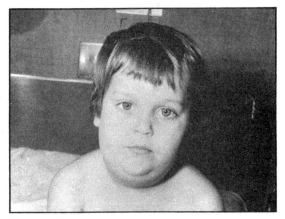

Figure F

The mumps is an infection of the salivary glands. Did you ever have the mumps?

At one time many children had the mumps. Now fewer children have this infection because they have been given a vaccine against it.

THE SEARCH GOES ON

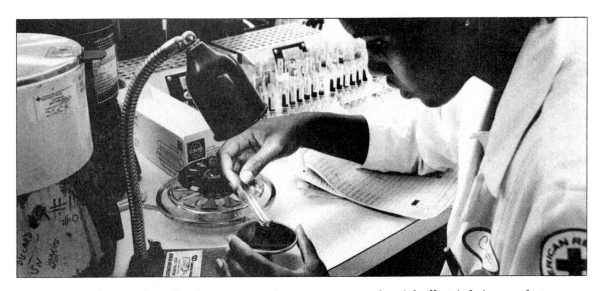

Figure G *The search to develop new vaccines never stops. Special effort is being made to develop vaccines against cancer and AIDS.*

TRUE OR FALSE

In the space provided, write "true" if the sentence is true. Write "false" if the sentence is false.

____False____ **1.** Acquired immunity is immunity that you are born with.

____True____ **2.** Vaccines are helpful in preventing polio and the measles.

____True____ **3.** Resistance to a certain disease is called immunity.

____True____ **4.** Being given a shot of antibodies is one way to acquire immunity.

____False____ **5.** There are vaccines for every disease.

____False____ **6.** Vaccines are given only by injection.

____False____ **7.** A vaccine causes you to get a disease.

____True____ **8.** A booster may be required to keep immunity against some diseases.

____True____ **9.** People are born with natural immunity.

____True____ **10.** Developing babies get antibodies from their mothers.

COMPLETE THE TABLE

Complete the table by identifying the kind of immunity that is described in the first column. Place a check mark in the correct column.

	Description	Natural Immunity	Acquired Immunity
1.	You are injected with a vaccine.		✔
2.	You are exposed to chicken pox.		✔
3.	You are born with an immunity.	✔	
4.	You are given a shot of antibodies.		✔

What are other ways of fighting disease?

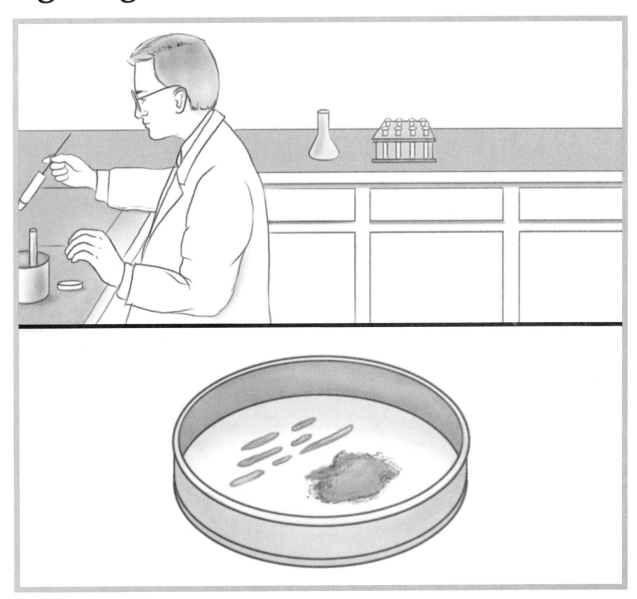

KEY TERM

antibiotic: chemical substances that kill harmful bacteria

LESSON 19 | What are other ways of fighting disease?

Your body has many defenses to protect itself from disease. But sometimes, the body needs help. At these times your doctor may prescribe medicine for you. The doctor may also give you an injection.

Have you ever been prescribed penicillin [pen-uh-SYL-in]? Penicillin was the first **antibiotic** [an-ti-by-AHT-ic] to be discovered. Antibiotics are chemical substances that kill harmful bacteria.

Penicillin was discovered in 1929 by Alexander Fleming. Fleming was an English scientist. He was growing bacteria in a dish and noticed that bacteria did not grow in one part of the dish—the part of the dish where some mold had grown. Fleming guessed that the mold produced a substance that was harmful to bacteria. He was right! The substance was penicillin. Penicillin destroys some bacteria and stops it from reproducing.

There are other antibiotics too. Like penicillin, most are made from molds. However, some come from bacteria and plants.

Antibiotics are not all the same. Each antibiotic can only be used to treat certain diseases. And no antibiotic works against viruses.

Before giving a person an antibiotic, most doctors ask if the person is allergic to it. That is because antibiotics cause allergic reactions in some people. They may get a fever or rash. In serious cases, a person may not be able to breathe. You should always tell your doctor if you are allergic to any medicines.

The discovery of disease-fighting medicines has been very important in treating disease. But the best way to "treat" disease is to prevent it. Here are some ways you can help prevent disease.

Many infectious diseases can be prevented by proper sanitation.

Figure A

Proper canning, pasteurization [PAS-chuh-ruh-ZAY-shun], and refrigeration help prevent infectious diseases caused by food poisoning.

Figure B

Perhaps the best way to avoid disease is to live a healthy lifestyle. People often get sick when their bodies are run-down. And doctors think that living in a healthy way lessens your chance of heart disease and some types of cancer.

How can you live a healthy lifestyle?
- Exercise on a regular basis.
- Eat a balanced diet.
- Get enough rest.
- Avoid harmful substances like tobacco.

Figure C

TRUE OR FALSE

In the space provided, write "true" if the sentence is true. Write "false" if the sentence is false.

__True__ 1. Penicillin is made by a mold.

__False__ 2. Antibiotics are all the same.

__False__ 3. Most antibiotics are produced by plants.

__True__ 4. Proper sanitation can help prevent disease.

__True__ 5. Some antibiotics are made by bacteria.

__False__ 6. Penicillin works against viruses.

__True__ 7. An antibiotic is a chemical substance.

__False__ 8. Living a healthy lifestyle has no affect on disease.

__True__ 9. Some people have allergic reactions to penicillin.

__True__ 10. Alexander Fleming observed that a mold produced a substance that was harmful to bacteria.

WORD SCRAMBLE

Below are several words you have used in this Lesson. Unscramble the words and write your answers in the spaces provided.

1. EMIDCIEN   MEDICINE

2. MDLOS   MOLDS

3. RATAEBCI   BACTERIA

4. CXEERIES   EXERCISE

5. CENLPILIIN   PENICILLIN

What is ecology?

KEY TERMS

biosphere: thin zone of the earth that supports all life

ecology: study of the relationship between living things and their environment

ecosystem: all the living things and nonliving parts of an environment

community: all the organisms living in a certain area

population: all the members of one species that live in the same area

LESSON 20 | What is ecology?

Our planet is huge. It has an area of more than 500 million square kilometers (200 million square miles). Yet life exists only on its surface, and slightly above and below. We call this narrow zone of life the **biosphere** [BY-uh-sfir]. You may know that the term "bio" means life.

The biosphere is full with all kinds of life. These organisms live in all kinds of environments. Everything that surrounds an organism makes up its environment. Living things are affected by their environment. They can also have an effect on their environments.

The study of the relationship between living things and their environment is called **ecology** [ee-KAHL-uh-jee]. Scientists who study ecology are called ecologists. The living and nonliving parts of a specific environment make up an **ecosystem** [EE-koh-sis-tum]. Some of the nonliving parts of an ecosystem are air, water, sunlight, and soil. Living things need these things to survive.

An ecosystem can be large, like an ocean or jungle. Or it can be small, like a pond or a patch of grass in an empty lot. Even a home aquarium is an ecosystem!

Each ecosystem is made up of one or more **communities**. A community is all the organisms living in a certain area. For example, a pond community may include frogs, fishes, and water lilies.

Members of a community depend upon each other. They also depend upon nonliving things like air, light, and water. The living and nonliving parts of the environment are always interacting. And a change in one part can cause a change in all the parts.

Each community is made up of **populations**. A population is all of the living things of the same species living in the same area. How many students make up the population of your class?

Figure A shows a lake ecosystem. The parts of this ecosystem are listed below. Next to each part, write <u>living</u>, if it is living. Write <u>nonliving</u>, if the part is not living.

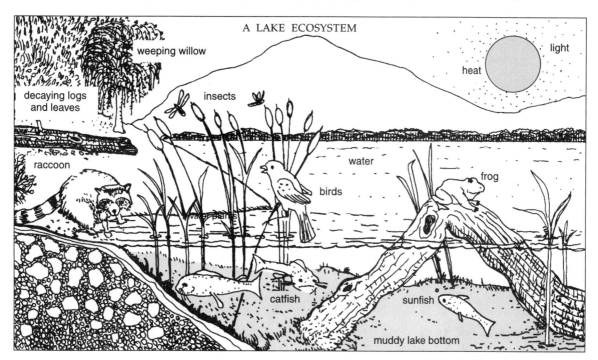

A LAKE ECOSYSTEM

Figure A *A lake ecosystem*

1. sunlight _____nonliving_____

2. catfish _____living_____

3. weeping willow tree

   _____living_____

4. raccoon _____living_____

5. heat _____nonliving_____

6. water _____nonliving_____

7. sunfish _____living_____

8. water plants

   _____living_____

9. frog _____living_____

10. muddy lake bottom

    _____nonliving_____

11. air _____nonliving_____

12. insects _____living_____

13. bird _____living_____

14. bacteria, algae, and other one-celled organisms (not shown, but always present in a lake ecosystem)

    _____living_____

15. Why are the one-celled organisms not shown? _____They are microscopic._____

Complete the following sentences.

1. An ecosystem is made up of _____ both living and nonliving _____ things.

2. All the living members of an ecosystem make up _____ a community _____.

3. The region of Earth where life exists is called the _____ biosphere _____.

4. All of the living and nonliving parts of an organism's surroundings are called its

    _____ environment _____.

5. Do living things affect nonliving things? _____ yes _____

    yes, no

6. Do nonliving things affect living things? _____ yes _____

    yes, no

7. A change in one part of an environment _____ can _____ cause a change in
    another part of the environment.

    can, cannot

8. The study of the relationship between organisms and their environment is called

    _____ ecology _____.

An aquarium is an ecosystem you may have in your home. A balanced aquarium is a healthy ecosystem. It is one in which all the organisms receive all the things they need to live.

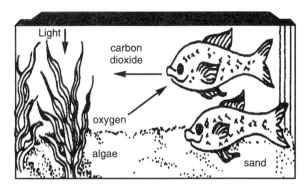

Figure B

What are the living and nonliving parts of an aquarium ecosystem?

living _____ fish, plants, algae _____

nonliving _____ water, carbon dioxide, _____

oxygen, sand, light

What are some other characteristics of an ecosystem?

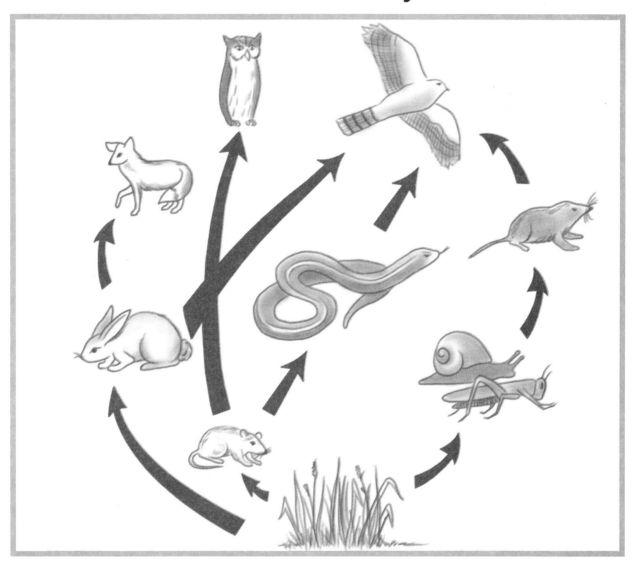

KEY TERMS

habitat: place where an organism lives

niche: an organism's role, or job, in its environment

producers: organisms that can make their own food

consumers: organisms that get food by eating other organisms

decomposers: organisms that feed on dead organisms

LESSON 21 | What are some other characteristics of an ecosystem?

If someone asked you where you live, how would you answer? The place where an organism lives is its **habitat** [HAB-i-tat]. A habitat is a special place. It provides <u>all</u> of an organism's needs, like food and air. It provides an organism with shelter. It also provides a place to reproduce. Sometimes, different species share the same habitat. For example, insects and mushrooms may share the same rotting log. Birds, squirrels, and insects might live in the same tree.

Now suppose someone asked what your role or job in life is. You would probably say that you are a student. Being a student is the job or role that you do where you live. Organisms also have jobs and roles in their communities. The job of a living thing is called its **niche** [NICH].

Living things may have the same habitat but they do not have the same niche. For example, tigers and deer both share a habitat in Asia. But while tigers chase and eat deer—deer eat grasses. They do not have the same role.

Although the tigers and deer in Asia have different roles, they <u>are</u> related by how they get their food. Each ecosystem is made up of different kinds of organisms.

Some are **producers**. Producers can make their own food. On land, the main producers are plants. In lakes and oceans, algae are the main producers.

Others are **consumers** [kun-SOO-murs]. Consumers get food by eating other organisms. Some consumers eat only plants. Others eat meat, or other animals. And some, like you, eat both plants and animals.

Some animals feed upon dead animals. They eat animals that have died or that have been killed by other animals. For example, vultures eat dead animals.

Bacteria break down the wastes or remains of organisms. They are **decomposers** [dee-kum-POHZ-ers]. Decomposers return materials from dead organisms to the soil.

Living things depend upon each other for food. Every living thing is a link in a <u>food chain</u>. A food chain shows the order in which living things feed upon other living things.

Look at Figure A. It shows a food chain. The arrows in the food chain show the direction that food moves along the chain.

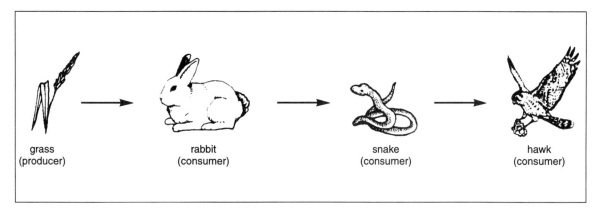

grass
(producer)

rabbit
(consumer)

snake
(consumer)

hawk
(consumer)

Figure A

Not all organisms eat the same kinds of food. Therefore, there are many different food chains. But, all food chains begin with <u>PRODUCERS</u>.

WHY?

Producers are the only organisms that can make their own food, using energy from the sun.

Why is the sun the source of energy in an ecosystem? ___All organisms get their food___

___either directly or indirectly from sunlight; food is an organism's energy source.___

_____

_____

Six food chains are shown below. One link has been left out of each chain. Identify the organism that is missing. Write your answers in the proper spaces below. Some blank spaces have more than one answer.

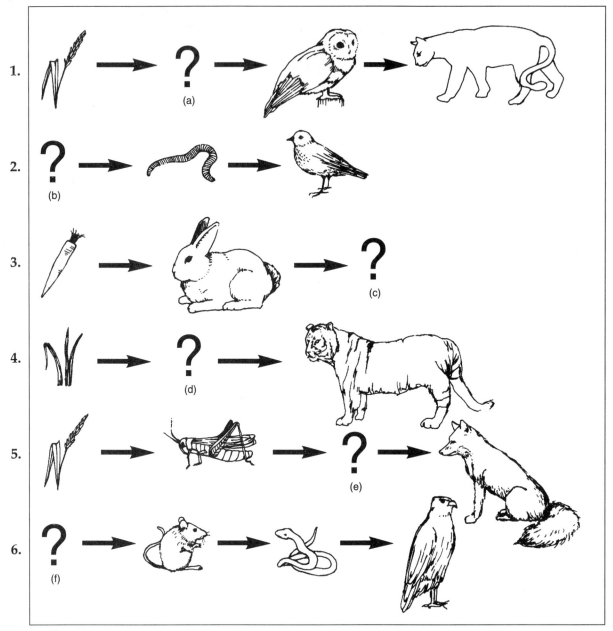

Figure B

1.	**a.**	accept any plant-eating consumer	**4.** **d.** accept any plant-eating consumer
2.	**b.**	accept any producer	**5.** **e.** accept any meat-eating consumer
3.	**c.**	accept any meat-eating consumer	**6.** **f.** accept any producer

136

You have just learned that food chains show food relationships. However, in nature, many food chains combine and overlap. They form a <u>food web</u>. A food web is a more complete way of showing food relationships. A food web shows how a number of food chains are related.

Look at the food web in Figure C. Then answer the questions.

1. What is the diagram shown called?

   food web

2. What does the diagram show?

   how food chains are related

3. What does a rabbit eat? __grass__

4. What organisms do wolves eat?

   rabbits and birds

5. Which organism is the producer?

   wheat

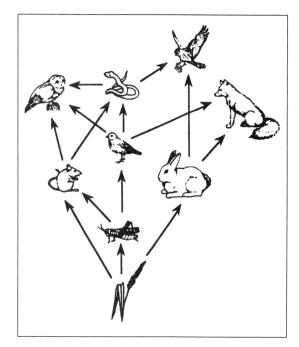

Figure C

MAKE YOUR OWN FOOD CHAIN

In the space provided, draw one of the food chains shown in the diagram above.

Accept any logical chain.

*Classify each description as a **habitat** or **niche** by checking the correct column.*

Habitat and Niche			
	Description	**Habitat**	**Niche**
1.	Eaten by fish		✔
2.	Under rocks	✔	
3.	Hole in a tree	✔	
4.	Eat mice		✔
5.	Nest on a tree branch	✔	
6.	Eat seeds and fruit		✔
7.	Log	✔	
8.	Jungle	✔	
9.	Shared by organisms	✔	
10.	Not shared by organisms		✔

MATCHING

Match each term in Column A with its description in Column B. Write the correct letter in the space provided.

Column A

_____e_____ **1.** plants

_____a_____ **2.** producer

_____d_____ **3.** decomposer

_____b_____ **4.** consumer

_____c_____ **5.** vulture

_____f_____ **6.** algae

Column B

a) organism that makes its own food

b) animal that feeds on other animals

c) bird that eats dead animals

d) organism that breaks down the wastes or remains of other organisms

e) main producers on land

f) main producers in lakes and oceans

Classify each organism listed in the table as a *producer, consumer, or decomposer. Place a check mark in the correct column.*

	Organism	**Producer**	Consumer	Decomposer
1.	Seaweed	✔		
2.	Duck		✔	
3.	Hawk		✔	
4.	Ants		✔	
5.	Bacteria			✔
6.	People		✔	
7.	Rabbits		✔	
8.	Grass	✔		
9.	Apple Tree	✔		
10.	Bees		✔	
11.	Earthworm		✔	
12.	Beetle		✔	

Complete each statement using a term or terms from the list below. Write your answers in the spaces provided.

niche   webs   food
sun   soil

1. A producer can make its own _____food_____ .

2. The _____sun_____ is the **source of energy** for an ecosystem.

3. **Food chains combine to form food** _____webs_____ .

4. **The role of an organism is called its** _____niche_____ .

5. **A decomposer returns materials from dead organisms to the** _____soil_____ .

Use the clues to complete the crossword puzzle.

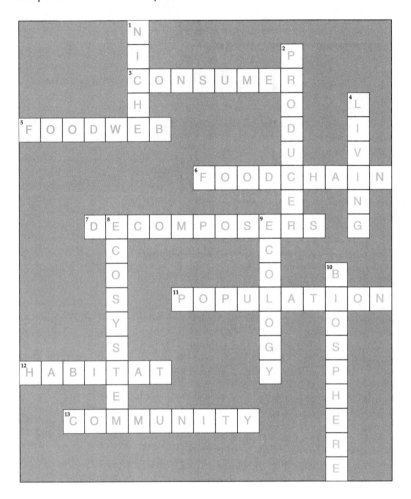

CLUES

ACROSS

3. organism that gets food by eating other organisms

5. combining and overlapping of many food chains

6. model of the flow of energy through an ecosystem

7. organism that feeds on dead organisms

11. all the members of one species that live in the same area

12. place where an organism lives

13. all the organisms living in a certain area

DOWN

1. an organism's role in its environment

2. organism that makes its own food

4. not dead

8. all the living and nonliving parts of an environment

9. study of the relationship between living things and their environment

10. thin zone of the earth that supports all life

What are biomes?

KEY TERM

biomes: large region of the earth that has characteristic kinds of organisms

LESSON 22 | What are biomes?

The **biosphere is** divided into major areas called **biomes** [BY-ohms]. A biome **is determin**ed mainly by its climate-like temperature and rainfall. Each **biome has a** different climate. Climate, in turn, affects the soil. The earth's **land areas** are divided into six major biomes. They are:

TUNDRA **Most** of the year, the tundra is bitterly cold and covered with **snow and ice.** The ground remains permanently frozen. It is called **permafrost. Only** certain small plants such as mosses and grasses can grow **in the tund**ra. Some animals, like reindeer and foxes, move in during **the grow**ing season. But they move out again as the frigid weat**her approaches.** Very few animals live year-round in the tundra.

CONIFEROUS FOREST Conifers are cone-bearing trees such as pines and fir **trees. Conifers** make up the coniferous forest biome. It is an area with a **cold climate.** Conifers form dense forests. The tree tops block out much **of the sunlight.** Grasses and smaller trees cannot grow. Only some shrubs**, ferns, and** mosses thrive. Coniferous forests are "home" for many **animals, su**ch as squirrels, moose, birds, and insects.

DECIDUOUS FOREST Deciduous trees such as maples and oaks shed their **leaves in the** fall. Deciduous forests thrive in moderate climates. Summe**rs may be** hot and winters may be cold. But temperatures do not get too **hot or too** cold for a very long time. Deciduous forests receive a good **supply of** water. They form dense forests. A deciduous forest provid**es habitats** for many kinds of animals.

TROPICAL RAIN FOREST A tropical rain forest is very warm and very **moist all the** time. It receives plenty of sunlight and rain. This environ**ment is ex**cellent for plant growth and soil development. Plants grow **thick and ta**ll. Tropical rain forests are found in areas near the equato**r. Rain fores**ts support more plant life and animal species than any other **biome.**

GRASSLANDS The chief plant life in the grassland is grass. Grassland and **deciduous** forest temperatures are about the same. But grasslands do not **receive as** much rainfall. Grasslands get enough rain to support grasses**—but not** trees. Grasslands are excellent for grazing animals. The soil **of grass areas** is very rich. Wheat and corn are grown here. Grasslands **also are** "home" for many small, burrowing animals.

DESERT A **desert** biome is very dry. It receives very little rainfall. Deserts **are very h**ot during the day, but they are cold at night. Desert soil is **very dry and** poor. Because of this, only a few kinds of plants grow in the **desert. And,** very few animals can survive in the desert.

The map below shows the major land biomes of the earth.

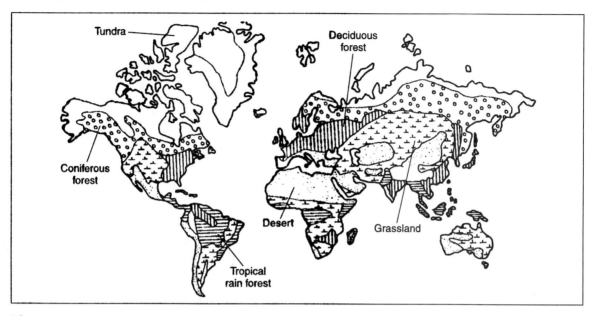

Figure A

1. **In which biome do you live?** _____Answers will vary._____

The chart below shows the climates of the major land biomes. Study the chart and then answer the questions.

BIOME	AVERAGE YEARLY RAINFALL	AVERAGE YEARLY TEMPERATURE RANGE
Tundra	less than 25 cm (10 in.)	-25°C–4°C (-13°F–39°F)
Coniferous forest	35–75 cm (15–30 in.)	-10°C–14°C (14°F–57°F)
Deciduous forest	75–125 cm (30–50 in.)	-6°C–28°C (43°F–83°F)
Tropical rain forest	200–450 cm (80–175 in.)	25°C–28°C (77°F–83°F)
Grassland	25–75 cm (10–30 in.)	0°C–25°C (32°F–77°F)
Desert	less than 25 cm (10 in.)	24°C–40°C (75°F–104°F)

2. **What is the average yearly temperature range of the tropical rain forest biome?**

   _____25°C–28°C (77°F–83°F)_____

3. **What biome gets between 75 and 125 cm of rainfall per year?** _____deciduous forest_____

The photographs show the six major land biomes. Identify each biome. Write the name of the correct biome on the line below each photograph.

Figure B

1. This photograph shows a

   _____desert_____ biome.

Figure C

2. This photograph shows a

   _____coniferous forest_____ biome.

Figure D

3. This photograph shows a

   _____tropical rain forest_____ biome.

Figure E

4. This photograph shows a

   _____grassland_____ biome.

144

Figure F

Figure G

5. This photograph shows a

   _____tundra_____ biome.

6. This photograph shows a

   _____deciduous forest_____ biome.

MULTIPLE CHOICE

In the space provided, write the letter of the word that best completes each statement.

____b____ **1.** Permafrost occurs in

    a) deserts.   **b)** the tundra.

    c) coniferous forests.   **d)** tropical rain forests.

____c____ **2.** Trees such as pines and firs make up

    a) tropical rain forests.   **b)** the tundra.

    c) coniferous forests.   **d)** grasslands.

____a____ **3.** The biome which supports more plant and animals species than any other is the

    a) tropical rain forest.   **b)** deciduous forest.

    c) coniferous forest.   **d)** grasslands.

____b____ **4.** Very few animals can survive in

    a) tropical rain forests.   **b)** deserts.

    c) grasslands.   **d)** deciduous forests.

____d____ **5.** Trees that shed their leaves in the fall make up

    a) the tundra.   **b)** coniferous forests.

    c) grasslands.   **d)** deciduous forests.

COMPLETE THE CHART

Study the characteristics of land biomes in the chart below. Complete the chart by placing a check mark in the correct column.

Land Biomes							
	Characteristics	**Tundra**	**Coniferous forest**	**Desert**	**Deciduous forest**	**Grassland**	**Tropical rain forest**
1.	Very hot days and very cool nights			✔			
2.	Trees with needle shaped leaves grow		✔				
3.	Used as farmland					✔	
4.	Hot and wet all year						✔
5.	Permafrost	✔					
6.	Maple and oak trees grow				✔		
7.	Cacti grow			✔			
8.	Spruce and moose are common		✔				
9.	Wheat and corn grow					✔	
10.	Jungles						✔
11.	Trees lose leaves in fall				✔		
12.	Conifers grow		✔				
13.	Reindeer live	✔					

What things can change the environment?

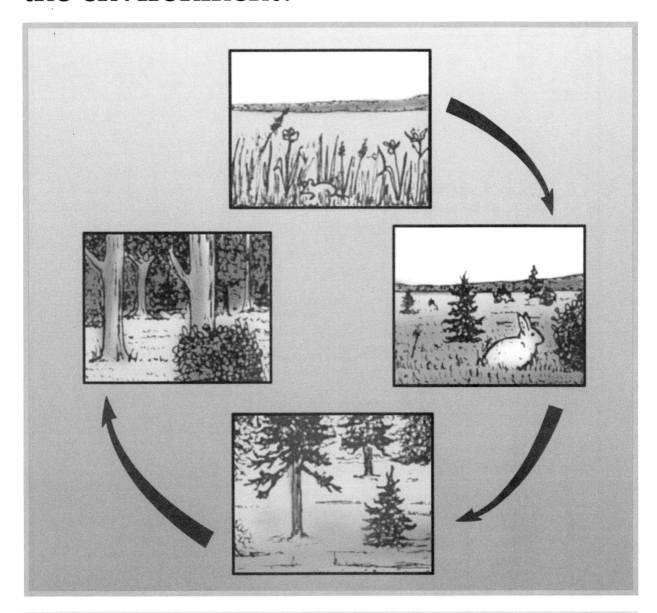

KEY TERM

succession: process by which populations in an ecosystem **are replaced** by new populations

LESSON 23 | What things can change the environment?

One of the world's greatest disasters took place on August 27, 1883. A volcano on the island of Krakatoa exploded. Much of the island was blown to bits. One part that had stood almost a kilometer high was left covered by nearly 275 meters of water.

The explosion caused a huge tidal wave. It swept over nearby islands. More than 36,000 people drowned.

The Krakatoa disaster caused great changes in the environment. Volcanic dust soared high into the atmosphere. Much of the sun's energy was blocked. Winds carried this dust around the world for more than a year. Temperatures dropped. Crops did not grow well. Animals were confused. They could not tell day from night.

Earth's history is a history of change. Some changes, like the Krakatoa volcano, earthquakes, lightning-caused fires, severe storms, floods, and droughts are natural events. Events change the environment. When the environment changes, its populations are slowly replaced by new populations. This process is called **succession** [suk-SESH-un].

A change in one group of organisms causes a change in another group. Changes first occur in plant populations. Then different animals move in.

Each year, more than four million acres of American forests are destroyed by fire.

Most of the plant life is destroyed. Many animals die; others flee.

Figure A

Nothing is left but ashes and black skeletons of what were once living trees.

The forest community is gone . . . But it will not stay that way. Many changes will take place to restore the forest. But it will take many years . . .

Figure B

1. First, grasses and weeds grow. They grow from roots and seeds left in the soil. They grow well. There are no trees to block the sunlight.

Figure C

2. These plants mature and form **seeds**. The wind spreads the seeds. Soon, a meadow forms. Small animals, **like** insects and birds, return to the **area**.

Figure D

3. Many growing seasons pass. **The** weeds, grasses, and insects add **minerals** to the soil. The soil **becomes** richer.

Figure E

4. The soil can support shrubs **and** small, fast-growing woody **trees**. These plants block the sun from **the** grasses and weeds. Other plants, like ferns, do not need full sunlight. **They** grow where the grasses and **weeds** once were. Different kinds of **animals** move in.

Figure F

5. The soil becomes richer. Taller, **slower**-growing hardwood trees grow. **Other** animals move in, such as rabbits, chipmunks, squirrels, and deer.

The area is now fully developed. **It is** a forest again. The community **will** remain in the area until the environment changes again.

Figure G

Complete each statement using a term or terms from the list below. **Write your** *answers in the spaces provided. Some terms may be used more than once.*

trees	changing	**succession**
shrubs	animals	**hardwood** trees
plant	environment	**grasses**
natural	weeds	

1. The earth is always _____changing_____ .

2. A slow change in populations of organisms in an area is **called** _____succession_____ .

3. In succession, the first changes occur in _____plant_____ **populations**.

4. If a forest burns down, _____grasses_____ and _____weeds_____ are the first to grow.

5. When plant populations change, different _____animals_____ **move in**.

6. Volcanoes and earthquakes are _____natural_____ events **that cause** changes.

7. In the final stage of succession, _____hardwood trees_____ **grow.**

8. As seasons pass, grasses and weeds make the soil richer, **allowing** _____shrubs_____

   and fast-growing trees to grow.

9. A community will remain in an area until the _____environment_____ changes.

10. Grasses and weeds grow well when there are no _____trees_____ **or**

   _____shrubs_____ to block the sunlight.

The steps below describe the destruction and rebuilding of a forest ecosystem. Place the steps in the proper order.

- shrubs and fast-growing short trees
- chipmunks and rabbits
- meadow
- dead forest
- grasses and weeds

- small birds and insects
- forest
- fire
- slow-growing, tall hardwood trees

1. ___ fire
2. ___ dead forest
3. ___ grasses and weeds
4. ___ meadow
5. ___ small birds and insects
6. ___ shrubs and fast-growing short trees
7. ___ slow-growing tall hardwood trees
8. ___ chipmunks and rabbits
9. ___ forest

How do people upset the balance of nature?

KEY TERMS

pollution: anything that harms the environment

pollutants: harmful substances

LESSON 24 | How do people upset the balance of nature?

An environment is constantly **changing. Sometimes, the changes work** together to keep the environment **in balance. In a balanced environment,** the size of the population **remains about the** same over time.

Sometimes the balance in an **environment is** upset. Many **times people** upset the balance of nature. **People upset the** balance **of nature by** destroying the habitats of oth**er living things. For example, people cut** down forests for farms and **towns. They build** dams **and dig mines. All** of these human activities **can be harmful** to other organisms in the environment. Many species **of animals are** finding it hard **to survive** because of the ways people **have upset the** balance of nature.

People also upset the balan**ce of nature** by causing **pollution [puh-** LOO-shun]. You probably **know that pollution** is a **major problem.** Pollution is <u>anything</u> that **harms the environment. It occurs when** harmful substances, or **pollutants [puh-LOOT-ents] are released into** the environment. Pollution **of the air,** land, and water are all **major** problems. Today many **different substances** are poisoning **the envi-** ronment and upsetting nature**'s balance.** And we cannot **think of just** air pollution, or <u>just</u> water **pollution, or** <u>just</u> land pollution. **Pollution** may <u>start out</u> in one part of **our environment.** <u>But it does not remain</u> <u>there.</u> It S-P-R-E-A-D-S to all **parts.**

Pollution is increasing daily. **Like other** organisms, people also suffer from the effects of pollution—**in the form** of illness, **birth defects,** respiratory diseases, and **many other problems.** Therefore, we **must all** work together to help reduce **pollution.**

Study **the pictures below** and read **the text desc**ribing each picture. Then answer the questions.

Figure A

The burning of fossil fuels is the major **cause** of air pollution. Oil, coal, and natural **gas** are fossil fuels. When these fuels **are** burned, many harmful substances are **rele**ased into the air.

1. How do you think car-pooling helps

reduce air pollution? ____It reduces____

the amount of fossil fuels burned.

_____

Figure B

When some harmful gases are released **into** the air, they combine with water to **form** acids. The acids fall to the earth as **acid rain.** Acid rain kills living things. It **also** damages buildings and statues.

2. What is acid rain? ____combination of____

harmful gases and water

_____

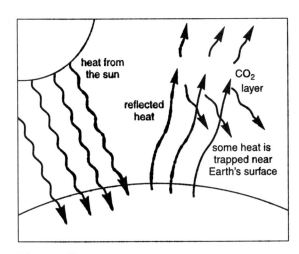

Figure C

Fuels need oxygen when they burn. They **give** off carbon dioxide. Carbon dioxide **traps** heat energy from the sun.

3. Scientists think that the increase of carbon dioxide in the air is causing the temperature of the earth to

_____rise_____.
                    rise, fall

155

Water pollution occurs when harmful substances enter the water. Major sources of water pollution include sewage, chemical wastes from factories, and fertilizers [FUR-tul-y-zuhrs], and pesticides [PES-tuh-sides] washed off farm fields.

You have probably seen cans, bottles, and papers thrown on the ground. These thrown-away materials are called <u>litter</u>. Litter is one cause of land pollution. Garbage and chemical wastes are other sources of land pollution. We produce billions of tons of garbage each year. Chemical wastes are often buried in the ground.

NOW TRY THIS!

Ten pollutants and pollutant sources are listed below.

Write **water** next to those that start out as water pollution. Write **land** next to those that start out as land pollution.

water	**1.**	raw sewage discharge
water	**2.**	dumping of chemicals into rivers
land	**3.**	pesticides
water	**4.**	detergents
land	**5.**	garbage
land	**6.**	burial of drums of toxic wastes
land	**7.**	fertilizers
land	**8.**	use of DDT to control mosquitos
land	**9.**	litter
land	**10.**	abandoning junk cars

Explain how a pollutant <u>gas</u> in the <u>atmosphere</u> can become

a) water pollution ___Answers will vary. Accept all logical responses.___

_____

b) land pollution ___Answers will vary. Accept all logical responses.___

_____

Complete each statement using a term or terms from the list below. Write your answers in the spaces provided.

spreads	illness	air
pollution	balanced	reduce
sun	does not	defects
sewage	survive	wastes
pollutants		

1. Anything that harms the environment is _____pollution_____.

2. Pollution occurs when _____pollutants_____ enter the environment.

3. The size of a population remains about the same in a _____balanced_____ environment.

4. Many species are finding it hard to _____survive_____ because of human activity.

5. Major sources of water pollution include _____sewage_____ and chemical _____wastes_____.

6. Pollution can cause _____illness_____ and birth _____defects_____ in people.

7. Pollution _____does not_____ stay in one place. It _____spreads_____ to all parts of the environment.

8. To help nature maintain a proper balance, we must _____reduce_____ pollution.

9. The burning of fossil fuels is the major cause of _____air_____ pollution.

10. Carbon dioxide traps energy from the _____sun_____.

REACHING OUT

About 66,000 square miles of the world's tropical rain forests are being destroyed each year. How does this upset the balance of nature? _____Answers will vary. Accept all logical responses.____

SCIENCE *EXTRA*

Wildlife Refuge Manager

How would you go about saving animals and plants? You could become a wildlife refuge manager. A wildlife refuge manager supervises a protected area for wild plants and animals. Refuges may be in forests, deserts, wetlands, or other habitats.

In order to properly protect plants and animals, refuge managers need to understand how the weather and soil affect living things. So they often measure temperature, rainfall, and soil chemistry. Refuge managers also work with the plants and animals directly. They count the numbers of different species living in the refuge. This is called taking a census. It is important for managers to know the population sizes of wildlife species they are trying to protect. Sometimes, however, there are so many plants or animals of one kind that it is impossible to count them all. In this case, managers estimate population size from a sample. This means that they count individual plants or animals in a small area, then compare the sample area to the size of the refuge.

Sometimes, refuge managers take active steps to protect a rare plant or animal. If a refuge contains a rare species of a bird, for example, the manager might keep visitors away from all nesting places, or close the refuge during the nesting season.

You need a bachelor's degree in biology, botany, or zoology to become a wildlife refuge manager. Preparing reports and analyzing data from the field also are part of the manager's job. So it is helpful to study mathematics, statistics, and technical writing as preparation for this career. You should also enjoy outdoor work, including working in the rain, snow, heat, or cold. Refuge managers sometimes must be out in bad weather, because protecting wildlife cannot always wait for a warm sunny day.

What is conservation?

PAPER **GLASS** **ALUMINUM**

National
Wildlife
Refuge

KEY TERMS

conservation: wise use of natural resources

natural resources: materials and energy in the biosphere that are used by living things

renewable resources: resources that can be replaced by nature

nonrenewable resources: resources that cannot be replaced

LESSON 25 | What is conservation?

Think about your pantry. It contains many kinds of foods. When they run low, you replace them. But what if you knew that you could not replace certain items? What would you do?

You would try to make them last as long as possible. You would use them sparingly, or conserve them. Conserve means to protect from being used up.

Earth is like a huge pantry. It has all the things we need to stay alive. It is also stocked with things that modern people use, like ores for metals and fuels for energy. All of the things we get from the environment are called **natural resources**.

There are two main groups of natural resources, renewable resources and nonrenewable resources.

RENEWABLE RESOURCES are replaced by nature. Oxygen, water, soil, and living things are renewable resources. Oxygen is made by plants during photosynthesis. Soil is made when rocks break up. Water is renewed through the water cycle. Living things reproduce themselves.

NONRENEWABLE RESOURCES are not replaced by nature, at least, not in a reasonable period of time. Fossil fuels, such as oil, coal and natural gas are nonrenewable resources; so are minerals. We get metals from mineral ores. How would your life be different without fossil fuels and metals?

At one time our supply of natural resources seemed endless. Now we know differently. The population of the world is increasing. We are using more, wasting more, and polluting more natural resources than we did in the past.

Earth's "pantry" is limited. We must use our resources wisely. If we do not, there will not be enough resources left for future generations.

The wise use of our natural resources is called **conservation** [kon-sur-VAY-shun]. Conservation of all natural resources, including renewable resources is important. Even though renewable resources are replaced, their supply is limited. People must be careful not to use them up faster than they can be replaced.

AIR CONSERVATION

Air pollution is very bad in most industrial areas and cities. But, air pollution spreads everywhere. It reaches every place on Earth.

Polluted air can smell bad. It can cause health problems, like respiratory diseases, lung cancer, and allergies. Polluted air also kills trees and reduces food crops.

Motor vehicles and factories are the major causes of air pollution. Strict laws concerning air pollution must be passed and enforced.

Figure A

What laws would you suggest to help cut down on air pollution? _____ Answers will vary.

Accept all logical responses.

One of the least expensive ways to control air pollution is to walk instead of drive and use public transportation.

WATER CONSERVATION

The average person drinks about 228 gallons of water each year. But water is not only used for drinking. We use it in many other ways. For example, we use water for bathing, swimming, cleaning, cooking, gardening, and boating. Water is also vital for proper sewage disposal, industry, agriculture, and aquaculture.

Our water supply must be kept fresh and safe to drink. We must stop dumping wastes and raw sewage into our water supply. We can also conserve water by turning off the water when brushing teeth and by taking showers instead of baths.

Figure B

What other ways can you conserve water? _____ Answers will vary. Accept all logical

responses.

It takes nature from 500 to 1,000 years to produce about two and one-half centimeters (one inch) of topsoil.

Soil can be carried away by wind and moving water. This removal of soil is called erosion. Erosion can be reduced.

To prevent too much soil erosion, people must practice soil conservation.

Figure C

Some ways to conserve soil are:

a) Cover the soil with plants such as grass or shrubs. The roots of plants help hold soil together.

b) Restrict the cutting down of forests. Trees act as wind breaks and help prevent soil from being blown away by wind.

c) Plant crops across the slope of a hill instead of up and down the hill. This helps prevent soil from being carried away by water running down the hill.

d) Add materials like humus or natural fertilizers to the soil.

Figure D

All the natural plants and animals living in an area are called **wildlife**. Wildlife are part of nature. They provide us with food, clothes, and many other products. Wildlife are also pretty to look at.

Figure E

Human activity can cause many forms of wildlife to become extinct (die out). We pollute and overhunt. We destroy wildlife habitats for construction and mining. These actions disturb nature's balance.

Some methods of wildlife conservation are:

a) Protecting the habitats of organisms.

b) Enforcing strict hunting and fishing laws.

c) Setting aside refuges, parks, and other public lands for wildlife.

d) Providing special "breeding grounds" for endangered (organisms in danger of becoming extinct) species.

Forests are "home" for many plants and animals. Forests provide us with oxygen, lumber, wood pulp, medicines, and many other products. Wood pulp is used to make paper, including the pages of this book.

Tropical rain forests have more plant and animal species than any other place on earth.

Many of the habitats being lost are in the tropical rain forests.

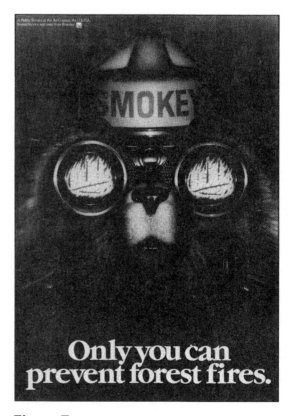

Figure F

Fires caused by human carelessness destroy many forests. Public education about dangers of forest fires is one way to help conserve forests.

Other methods of forest conservation include:

a) Planting new trees to replace those that have been chopped down for lumber or other products.

b) Chopping down only certain parts of a forest to allow seeds from remaining trees to provide replacements.

c) Removing only older or unhealthy trees from forest regions.

Figure G

METAL CONSERVATION

To recycle means to "use over again." Some resources, like metals, can be recycled. Aluminum cans, glass bottles, newspapers and some of the metals used to make cars can all be recycled. They can be melted down and reused. Most can be recycled over and over again. Recycling is an important way to conserve minerals.

Figure H

Fuels are nonrenewable resources. Once a fuel is used, it is gone. It cannot be recycled. The best way to conserve a fuel is not to waste it. Use it sparingly. Use it as you would a nonreplaceable item in your pantry.

Here are some ways people can help conserve fuel.
a) Use cars that get good gas mileage. Drive within the speed limit. Don't speed.
b) Walk, bicycle, or car pool when possible.
c) Turn off lights when you leave a room. This saves the fuel used to produce electricity.
d) Use electrical appliances that are energy saving.

MULTIPLE CHOICE

In the space provided, write the letter of the word that best completes each statement.

_____c_____ **1.** All the things nature gives are called
    a) pantries   **b)** ores
    c) natural resources   **d)** renewable resources

_____a_____ **2.** The things that nature replaces in a short period of time are called
    a) renewable resources   **b)** nonrenewable resources
    c) fossil fuels   **d)** wildlife

_____c_____ **3.** An example of a renewable resource is
    a) coal   **b)** aluminum ore
    c) oxygen   **d)** oil

_____b_____ **4.** Things that nature does not replace in a reasonable period of time are called
    a) renewable resources   **b)** nonrenewable resources
    c) pollution   **d)** natural resources

_____d_____ **5.** An example of a nonrenewable resource is
    a) water   **b)** soil
    c) air   **d)** minerals

_____c_____ **6.** The wise use of our natural resources is called
    a) recycling   **b)** consideration
    c) conservation   **d)** erosion

_____a_____ **7.** Organisms in danger of dying off are considered
    a) endangered   **b)** extinct
    c) conserved   **d)** wildlife

_____c_____ **8.** The use of resources over and over again is called
    a) erosion   **b)** cycling
    c) recycling   **d)** replacement

MATCHING

*Match each term in Column A with its **description in Column B. Write the correct letter** in the space provided.*

	Column A		Column B
d	1. water, air, soil and living things	a)	wise use of resources
c	2. minerals and fossil fuels	b)	cause of most pollution
a	3. conservation	c)	nonrenewable resource
e	4. pollution	d)	renewable resource
b	5. people	e)	harms all living things

REACHING OUT

Make a list of five things you use often. **Identify the natural resource (or resources)** that each of the things you listed came from.

ITEM	NATURAL RESOURCE(S)
Lists will vary. Accept all logical answers.	

METRIC-ENGLISH CONVERSIONS

	Metric to English	English to Metric
Length	1 kilometer = 0.621 mile (mi)	1 mi = 1.61 km
	1 meter = 3.28 feet (ft)	1 ft = 0.305 m
	1 centimeter = 0.394 inch (in)	1 in = 2.54 cm
Area	1 square meter = 10.763 square feet	$1 ft^2 = 0.0929 m^2$
	1 square centimeter = 0.155 square inch	$1 in^2 = 6.452 cm^2$
Volume	1 cubic meter = 35.315 cubic feet	$1 ft^3 = 0.0283 m^3$
	1 cubic centimeter = 0.0610 cubic inches	$1 in^3 = 16.39 cm^3$
	1 liter = .2642 gallon (gal)	1 gal = 3.79 L
	1 liter = 1.06 quart (qt)	1 qt = 0.94 L
Mass	1 kilogram = 2.205 pound (lb)	1 lb = 0.4536 kg
	1 gram = 0.0353 ounce (oz)	1 oz = 28.35 g
Temperature	Celsius = 5/9 (°F–32)	Fahrenheit = 9/5°C + 32
	0°C = 32°F (Freezing point of water)	72°F = 22°C (Room temperature)
	100°C = 212°F	98.6°F = 37°C
	(Boiling point of water)	(Human body temperature)

METRIC UNITS

The basic unit is printed in capital letters.

Length	Symbol
Kilometer	km
METER	m
centimeter	cm
millimeter	mm

Area	Symbol
square kilometer	km^2
SQUARE METER	m^2
square millimeter	mm^2

Volume	Symbol
CUBIC METER	m^3
cubic millimeter	mm^3
liter	L
milliliter	mL

Mass	Symbol
KILOGRAM	kg
gram	g

Temperature	Symbol
degree Celsius	°C

SOME COMMON METRIC PREFIXES

Prefix		Meaning
micro-	=	0.000001, or 1/1,000,000
milli-	=	0.001, or 1/1,000
centi-	=	0.01, or 1/100
deci-	=	0.1, or 1/10
deka-	=	10
hecto-	=	100
kilo-	=	1,000
mega-	=	1,000,000

SOME METRIC RELATIONSHIPS

Unit	Relationship
kilometer	1 km = 1,000 m
meter	1 m = 100 cm
centimeter	1 cm = 10 mm
millimeter	1 mm = 0.1 cm
liter	1 L = 1,000 mL
milliliter	1 mL = 0.001 L
tonne	1 t = 1,000 kg
kilogram	1 kg = 1,000 g
gram	1 g = 1,000 mg
centigram	1 cg = 10 mg
milligram	1 mg = 0.001 g

GLOSSARY/INDEX

immunity [im-MYOON-i-tee]: resistance to a certain disease, 117

inbreeding: mating closely related organisms, 47

incomplete dominance: blending of traits carried by two or more different genes, 27

infectious [in-FEK-shus] disease: disease caused when a virus (or germs) enter the body, 101

mass selection: crossing organisms with desirable traits, 47

mimicry [MIM-ik-ree]: adaptation of an organism that protects the organism because its appearance is similar to another organism, 73

mucus [MYOU-kus]: sticky substance made by the body that traps germs, 117

natural resources: materials and energy in the biosphere that are used by living things, 159

natural selection: survival of organisms with favorable traits, 59

niche [NICH]: an organism's role, or job, in its environment, 133

noninfectious diseases: diseases that are not caused by germs and not spread from person to person, 109

nonrenewable resources: resources that cannot be replaced, 159

nucleic [new-KLEE-ik] acids: organic compounds that make proteins, control the cell, and determine heredity, 95

opposable thumb: a thumb that can touch all of the other fingers, 79

pollutants [puh-LOOT-ents]: harmful substances, 153

pollution [puh-LOO-shun]: anything that harms the environment, 153

population: all the members of one species that live in the same area, 129

primates: order of mammals, 79

producers: organisms that can make their own food, 133

Punnett square: chart used to show possible gene combinations, 21

pure: having two like genes, 15

recessive [ri-SES-iv] gene: weaker gene that is hidden when the dominant gene is present, 15

renewable resources: resources that can be replaced by nature, 159

sex chromosomes: X and Y chromosomes, 35

succession [suk-SESH-uh]: process by which populations in an ecosystem are replaced by new populations, 147

toxins: poisons, 102

traits: characteristics of living things, 1

vestigial [ves-TIJ-ee-uhl] structures: body parts that are reduced in size and that serve no function, 65

virus: piece of nucleic acid covered with an outercoat of protein, 95

white blood cells: cells that protect the body against disease, 117